The Nature of Summer

Jim Crumley

Saraband

Published by Saraband
Digital World Centre, 1 Lowry Plaza,
The Quays, Salford, M50 3UB
and
Suite 202, 98 Woodlands Road,
Glasgow, G3 6HB, Scotland
www.saraband.net

This paperback edition 2021

ISBN: 9781913393113
ebook: 9781912235735
Printed and bound in Great Britain by Clays Ltd, Elcograf S.p.A.

1 2 3 4 5 6 7 8 9 10

Contents

REMEMBER WELL

Neil Campbell MacArthur

1949–2019

Prologue

The Goddess of Small Things

Simmer's a pleasant time
Flowers of every colour
Water rins o'er the heugh
And I long for my true lover
Robert Burns

CONSIDER THE MOUNTAIN SORREL by your left boot. If you failed to spot it don't worry, you wouldn't be the first. At 4,000 feet on the Cairngorms plateau, there is bigger and more handsome stuff to look at. But summer is the Goddess of Small Things. So now that I have drawn your attention to it, why not give the mountain sorrel the time of day? I know, I know, it looks like nothing at all; it's basically a high-altitude dock leaf. What's this one…four inches tall? Yet up here, things have a habit of not quite looking like what they really are. That sparse cluster of kidney-shaped leaves at ground level is what botanists call a basal rosette, which is arguably too grandiose for what actually meets your eye. And those things at the other end of a skinny stem that morphs from red at the bottom to green halfway up, those are what pass for flowers, and it is true that they are on the nondescript side of insignificant. But let me show you something. Look closer, look deeper, look *inside* the flower. See the whole plant. The way to see what's there is to get down on your knees. Peel the petals apart. Do you see it? This is the fruit of the mountain sorrel, not a berry but a nut. I told you we were dealing with small things. It's about an eighth of an inch long. Three millimetres, if you don't do fractions. Turn your binoculars upside down, put the eyepiece almost against the nut and look in the wrong end, for now you have a microscope in your hand. And now that you can see it larger than life, what do you think that is, that green canopy to which the nut clings? Can you see how beautifully formed it is, like an open book; and can you see that it is exquisitely edged in red, the way the finest book pages are edged in gold? It's a wing. So when plateau winds blow (and the wind has a considerable repertoire up here, from the easiest of breezes like this July morning to an all-Britain all-time record of 176 miles per hour in January 1993), the nut flies until it eventually touches down and – in time, in time – it pushes a root into the tough plateau soil and a new mountain sorrel plant begins to come to terms with high living.

Now consider its neighbour. Notice that unlike the mountain sorrel's erect stem and spike of flowers and winged nuts, its neighbour is a horizontal, ground-level straggle of shining leaves. Such is the nature of summer in the high Cairngorms that ten days ago this strange growth showed not so much as a leaf bud. Plants of all kinds bloom late and wither early here. The growing season, such as it is, is fast and brief. These leaves are fully open. And if you care to lift up one or two, you may find a yellowish non-leaf growing among them, and if you have the capacity to set aside the evidence of your eyes and think outside the box, it may occur to you that it looks like a tiny catkin – because that is what it is. What you are looking at is a tree, an inch-high tree with its "branches" underground, a dwarf willow. And this is its Scottish homeland, the highest, pared-to-the-bone upthrusts of the Cairngorms, and what passes for summer up here is a short, sharp shock of a season (in the forty-something years I have known these mountains, I have acquired a complete snow calendar: that is, I have been snowed on in every month of the year, so including June, July and August). So short and so sharp that the leaves of some specimens have turned yellow in July, while others just a few hundred yards away are still in bud or have yet to bud at all.

The law of unintended consequences comes into play at this point, for the dwarf willow shares its homeland and its version of summer with ground-nesting dotterels. Ground-nesting, because up here there is nowhere else to nest. Commonplace words acquire a different meaning. "Nest" in the dotterel's case is the shallowest of shallow scrapes, a hint of a depression in the raw surface of the mountain; and "soil" is really not an appropriate word at all. The only possible explanation for the dotterel's choice of nest site is that its true heartland is the Arctic tundra, and the broad, bare plateau of the high Cairngorms is the nearest thing we have to an Arctic landscape.

Seton Gordon is still the supreme bard of this landscape almost a hundred years after he wrote its masterwork, *The*

Cairngorm Hills of Scotland (Cassell, 1925), and his account of his life-changing 1921 Oxford University expedition to Spitzbergen, *Amid Snowy Wastes* (Cassell, 1923). Forever after, he would draw direct comparison between the two landscapes and point out the similarities between the Cairngorms above 3,500 feet and Spitzbergen at sea level. Thus, much of the dotterel's fragile Scottish population of around 400 breeding males centres around the Cairngorms, and most nests are at or above 4,000 feet, whereas in the Arctic it finds what it is looking for at about 100 feet above sea level. In the high Cairngorms, it likes to line its nest with moss, lichen and (if it can find them) leaves. The easiest leaves to pluck from any hardwood tree are always the withering ones that have changed colour, and in the case of all willows that means from whitish green to yellow. So the yellow leaves of some of the autumn-minded dwarf willows even in July are not just the easiest to pick, they are also the easiest to see. As Seton Gordon wrote in 1925:

> *The old withered yellow leaves are used by the dotterel for lining her nest.*

I think it is quite possible that he was the first person to observe such a thing, certainly the first to think it was worth writing down.

◉ ◉ ◉

A trackside bank in Balquhidder, at an altitude 3,500 feet less than a Cairngorms dotterel nest, is where I find my first fragrant orchids. I like orchids, the little wild ones that gladden a Scottish summer, not the haute couture monsters beloved by interior designers. There are 7,500 species in the world, around fifty of which are found in Britain, and no more than a dozen in Scotland (not including a bewilderment of sub-species and hybrids which I choose not to care about). An orchid that smells of carnations is a little special – hence, fragrant orchid. In the sparsely populated landscape of Scottish orchids, then, fragrant orchids are particularly thin on the ground. But when you do find them, they are worth lingering over: they are small and discreet, the colour is soft, subtle, pale and pink, and the mild but distinct carnation scent is sensational. In the glitzy world of orchids, this is the one that gives understatement a good name, the less-is-more orchid. Ah, but if you look close enough, and if you are willing to linger long enough and get your eye in – *see the whole plant* – you will unearth the fragrant orchid's scintillating, shimmering secret. Take a flower spike very gently in your fingers, get your own shadow out of the way, and turn the spike a millimetre at a time in the sunlight. Keep turning and staring and refocussing until, finally, the miracle is revealed: every petal is covered in tiny glistening scales. As nature's light shows go, it is not outshone by either aurora or supermoon. The effect simply astounds, and so does the tiny nature of the spectacle. So when I suggest that summer is the Goddess of Small Things, these are what I have in mind, among other things.

Which other things? Oh, you know...

Goldcrest eggs. Especially second and third broods, around ten eggs at a time and about the size of the fingernail on my pinkie, or about half an inch, fledging well into the summer in a nest the size of a toddler's bunched fist – a nest fashioned from moss and lichen and bits of spider's webs and slung from the fronds at the outer edge of a Sitka spruce. In sunlight after rain it looks more like a silk purse than a nest. Unlike the human population of Scotland, the goldcrest loves Sitka spruce and thrives in it from Galloway to Sutherland. I am with the goldcrest; I like Sitka spruce. I don't like how we grow it and what we do with it, but I have seen it in the Alaskan Panhandle, where it grows with grace and elegance in the company of western hemlock, aspen, birch, willow, wolf, grizzly bear and even humpback whale (for the Pacific thrusts long, narrow, questing fingers into that forest so apparently land-locked that you think you are walking along the shore of a lake until a whale the size of a small island heaves out of the water and you remember where you are). Two other things you should know about Sitka spruce: when it is left to its own devices and grows wild, it produces timber of such quality that it is the choice of some of the finest luthiers in the world to make some of the finest guitars in the world, and even in the atrociously contemptuous way it is treated by contemporary forestry practice here, the goldcrest still can't get enough of it.

It is the unmistakable scent of red fox that stops me, that blend of spice and peat and something stuck to your boot. Just there, the forest track emerges from the old-established plantation into a high and open heather moor under a suddenly wide sky, the moor newly planted with more spruce but also studded with heather and rock and pine outcrops, where self-seeded spruce and a scatter of small rowans thicken an agreeable mix with a hint of Scandinavia.

Ben Ledi is suddenly there, as Ben Ledi so often is in this part of the world, and dark green and gold in the summer evening. At the base of the first of the outcrops, a hefty old pine had fallen years ago across a natural depression in the hillside. Time had homed in on the decomposing trunk and furred the walls and the floor of what had become a kind of accidental cave roofed in by the trunk and walled on one side by the massive, upended root. It was there that the foxes had denned, and in some comfort with very little effort on their part. It is a deep and dark and turned-aside place, and the only clue to the possibility of their presence is that cask-strength pungency. So you could say I have followed my nose. But just as I crouch towards the carefully discreet entrance, a Sitka spruce a few feet away unleashes a salvo of fluffy shrapnel. Never, I suspect, has a fresh-from-the-nest brood of goldcrests had such a startling impact on a six-feet-tall, thirteen-and-a-half stone mortal. A fledgling goldcrest is all of three inches from stem to stern, and as fluent in flight as a Kleenex tissue screwed up loosely into a ball. None of them flies more than a yard, a few have trouble perching, two try to get back into the nest, immediately betraying its presence within touching distance the moment I recover my balance and stand up. I would do better to stay crouched but the moment is unique in my long experience of keeping nature's company and I respond inelegantly. For perhaps ten seconds the air rocks with the whirring pulse of very small wings and there seem to be goldcrests everywhere. Then the noise stops and they vanish back into the spruce tree and are silent and still. I step quietly away from their tree. From the respectable distance of the forest track I apologise quietly for the intrusion. But now I know where they live, where the small miracle of them unfolds.

What other small things?

Dragonflies. Flanders Moss, a raised bog and national nature reserve a few miles west of Stirling, so more or less equidistant from east and west coasts, yet for all that, it lies at as near to sea level as makes very little difference. In early summer it transforms under the spectacular influence of an unbroken square mile of bog cotton, so dense and so level and so white that it effortlessly invokes snow. Summer snow is far from unknown in Scotland, and especially at 4,000 feet in the Cairngorms, but summer snow at sea level and the temperature at 20 degrees probably is pretty well unknown these last 10,000 years. No one notices the flowers of bog cotton. But after the flowers come the seed-heads, and each one of those is as the snowflake to the snowdrift. And it is this sea-level summer snow that is the setting for the dragonfly's finest hour, for both bog cotton and dragonfly need wet ground, preferably bogs, to thrive. Dragonfly: it is a confusing word, for it means both the species itself and the collective name for both dragonflies and damselflies. They inhabit the same landscape, and you will see them in sunlight. Page one of the Idiot's Guide will tell you that the main difference is how they perch: dragonflies perch with their wings open like an old bi-plane, and damselflies fold theirs neatly together along their bodies.

This is the story of one damselfly, or rather four. One species – the large red damselfly (which is large only in relation to the small red, two inches max) – but four members of that one species: two male and two female. How do I know? Because they are two mating pairs, and they are actually mating. I have a photograph. There is one pair, beautifully lit full-length, one behind the other on a bent-over blade of grass and low over a pond half-clothed in a species of moss with a Latin name I have no intention of learning. It makes a lovely backdrop to show off the banded red of the damselflies. And that is my shot. Having admired the photo several times (by my standards it is a good shot, which, admittedly, is to damn with faint praise), I have noticed that the female has only one of her six feet in contact with the grass, the rest are in the air. This could mean that she has been lifted bodily from the blade of grass, got carried away you might say; or it could mean that she was just touching down when I took the picture. I can't remember. But more noticeable by far now – although not at the time – is that the pair have been photo-bombed by a second pair, which I didn't notice as I pressed the shutter button because they are head-on rather than side-on, and a head-on pair of mating damselflies looks like nothing on earth you or I have ever seen before, and I am content to leave it at that.

I became interested in dragonflies/damselflies for two reasons. One is that I go to Flanders Moss often because it is close at hand, and it is at least four different places at different seasons of the year: nature plays many cards there. The other is that I met a man called Ruary Mackenzie Dodds, a relentless champion of the dragonfly/damselfly cause and the author of two books on the subject that are also published by Saraband. We shared a book festival gig at Wimborne in Dorset, as well as the flight from Edinburgh to Southampton and back and an agreeably boozy dinner. This kind of thing doesn't happen often, at least not to me, but we have become friends and I have become a dragonfly enthusiast by osmosis, as opposed to by incredibly hard work, which is how he did it.

Here is Ruary, writing in *The Dragonfly Diaries* (Saraband, 2014) on the subject of blue-tailed dragonflies:

> *...such lovely delicate things, with those powder blue blobs on the ends of their slim black abdomens...most have blue thoraxes but some have other colours – red, green, purple...these are females, refucens, infuscens and violacea respectively. Apparently the thorax colours can change with age...*

And here he writes on his wife Kari's obsession with dragonfly larvae rather than the flying beasties they become:

> *...they just don't have the same magic as the adults: no stunning flying, no stunning speed, no flashing beauty. There's that ferocious labial mask, though, and the way they breathe through their backside...*

Watching and listening to Ruary in full book festival flow is something of a roller-coaster ride, at the end of which you have been hugely entertained as well as informed, and you are just a little giddy and unsteady on your feet, and wondering how quickly you can get back to Flanders Moss and go and look for dragonflies again.

◉ ◉ ◉

Lizards, as well. Flanders Moss in summer is strewn with common lizards, and the chances are that you would never see one were it not for the boardwalk that allows you to watch the myriad life-forms of a raised bog at close quarters without getting your feet wet or floundering up to your waist in glaur, which is what nature uses to make peatbogs. The raised rim of the boardwalk is no more than three inches wide, and an inch higher than the main path. But the lizards home in on it to sunbathe. The star attractions are the newborn ones, perfect miniatures of the adults and an inch-and-a-half long, including the tail, but meticulously curved into half that length; two together look like parentheses. But move too suddenly and they dive headlong into the bog.

Flanders Moss, at first glance, is all about space and light and spectacle all the way to its rim of birchwoods and its arc of mountains beyond. But stop and stay still and look close, and it is the small things you will remember.

When I chose the title for this chapter, and having written it on a notebook page with a fountain pen (my preferred way of writing), I thought I would begin with a list. This was it:

Goldcrest eggs and nests and chicks, wild strawberries, wild raspberries, blueberries, brambles, cloudberries, cloudberry flowers, small blue butterflies, small coppers, small tortoiseshells, small whites, small pearl-bordered fritillaries, small skippers, dingy skippers, chequered skippers, orange-tips, northern-brown argus, Rannoch brindled beauty moth, sea pink, sea spurry, sea holly, newborn lizards, scales on the petals of fragrant orchids, small white orchids, seventeen species of speedwell, wild mountain thyme, mountain avens, mountain sorrel (and nuts), alpine lady's-mantle, dwarf cornel, eyebright, bedstraw, house martins, sand martins, sandpipers, wrens, merlins, little-ringed plovers, little auk, little tern, azure damselflies, all damselflies and dragonflies apart from those ones that look and sound like Sopwith Camels, spotted flycatchers, headdress of redpolls and reed buntings…

Then I ran out of ink, and I thought better of the idea and that perhaps you might just like to make your own list, now that I've shown you how to get the hang of it.

Part One

Everything Else in the Universe

Chapter One

St Kilda Summer, 1988

Behind Boreray

It is black behind Boreray and small.
All suns dance darkly here,
throw no shadows on rock this black.
Stac Lee is a black berg
its sunk seven-eighths beyond
the scope of suns and me.
We who sail our puny daring
under Stac an Armin
creep tinily by.

It all started in the early summer of 1988. I handed in my notice at the *Edinburgh Evening News*, where I had been working for eight years on the features desk, and at the invitation of publisher and landscape photographer Colin Baxter, I went to St Kilda, forty miles west of the Outer Hebrides, to write what would become my first book. Ian Nimmo, the editor of the *Evening News* at that time, and who had done much to encourage me to flex my writing muscles in his newspaper, gave me his blessing with the words "you must follow your star". So began an adventure that changed my life utterly. I had been a journalist from the age of sixteen. At the age of forty, I became a full-time nature writer literally overnight (and halved my income at a stroke). It was a fast transition: the book was published three months later, on the day after I left the paper. The only copy of it I still possess is the one I gave to my mother, and which I had inscribed:

For Mum with much love from the author!
Jim
October 1988

She had been very critical of my decision to leave the *Evening News* (she was not the only one among my family and friends and ex-colleagues). But seeing that book and holding it in her hands and reading the inscription…all that changed everything: her opinion of the enterprise swung through 180 degrees from a headwind to a tailwind, and for the last five years of her life until her death in 1993, she became the champion of my cause.

My first book? I can no longer remember how it felt. Probably I walked on air for a few days, then I looked around, thought, "What's next?" and started writing my second book. I thought that was how it was supposed to happen, and as no one has advised me differently, I have just kept on doing it ever since: forty books in thirty-two years. Still following my star, Ian. And thanks.

This book concludes a tetralogy of the seasons, and that life-redefining summer of 1988 seemed like a natural place to begin, the first of all my nature-writing summers, for all that it had been temporarily thwarted at the very first hurdle. On the day I was supposed to travel to Oban to join

Colin Baxter for the sail to St Kilda aboard a two-masted schooner, I was floored by a violent gastric bug. So I went alone three weeks later by the rather less glamorous route of a flight to Benbecula, in the Outer Hebrides, and then the Army's flat-bottomed landing-craft (the Army maintained a small base there at the time to service a cliff-top radar station). As one seasoned St Kilda veteran had counselled me: "She wallows like a drunken pig, that bitch."

As it happened, she declined to wallow. The evening ocean was as benign as the Crinan Canal. *HMS The Drunken Pig* was sober and demure. I stared at the ocean, at its raft of islands astern and its absence of islands ahead. I slept. Then a voice gatecrashed a dream: "Anybody want to see St Kilda? It's worth a look."

Oh, yes please. I wanted to see St Kilda very much indeed, for was it not to be the passport to the rest of my life? My watch said 5a.m. I went up on deck and the Atlantic was barely astir and the sky was pink and St Kilda was purple. And the voice was right: it *was* worth a look.

It is that first look that I remember, the one utterly indelible souvenir that has survived those thirty-two intervening years intact. It was a cardboard cut-out, as two-dimensional as a stage set, and it floated upright among leisurely waves. And it was purple. Or rather it was purples. The nearest island, a dour little tea-cosy-shaped rock lump called Levenish, was the darkest purple. My particular sightline set it against the much larger island of Boreray, and Boreray not only stood more than 1,200 feet straight up out of the ocean, it was the silhouette of a sea monster, and it was paler purple, borderline lilac. Over the next two weeks of camping alone, I would see Borerary from many angles in many weathers and every hour of the day and the dusk and dawn, and including the view through a gauze of gannets from the unhorizontal deck of a yacht at very close quarters indeed; but not once did that extravagant portfolio I would amass ever threaten to dislodge from my mind's eye that first of all my St Kildas on that first morning of all my nature-writing summers.

And Stac Lee was there too, a mere 600 feet high, but quite high enough for a lopsided parallelogram with no visible means of support, and that too was the paler shade of Boreray purple.

Then the boat dipped and I realised that parts of the superstructure were concealing parts of St Kilda, so I ran to the bow where wider oceanic miles lay unobscured. And there was Hirta, the main island, and there was the untidy sprawl of Dùn, Village Bay's eccentric, wafer-thin breakwater, looking like a bar of Toblerone that had gone horribly wrong in the baking.

These set pieces of the St Kilda archipelago, so familiar to me (in outline, at least) from a few books, from other people's photographs, from maps and film and drawings and paintings and word-of-mouth (it is astounding how many St Kilda veterans emerged like woodlice from under stones once the word of what I was doing leaked out) now made smithereens of my every preconception. While I stared and tried to respond in what I thought might be a suitably nature-writerly way to where I was and what I was seeing and what-the-hell-did-I-think-I-was-doing, by the way, something utterly new stole over me in the face of so much incomprehensible, volcanically tarnished age, and it was this:

Though I turned very slowly through 360 degrees, I could not see any other land, in any direction, none at all. And these first moments became my all-purpose visual definition of St Kilda, the one I have carried in my mind ever since. And so primitive was the encounter, so elementally simple – one sea, one sky, one scatter of improbable rocks – that I might have been the first of all St Kilda voyagers, one of a tribe of nomadic herdsmen coming curiously up the margins of Europe, exchanging bemused glances and agreeing among themselves that surely here was nature's last limit. For they would be accustomed to sailing where land was in sight, and on St Kilda, the only land very occasionally in sight (I would have one glimpse of a white-sanded Hebridean beach – Harris – in two weeks) is the land that you left forty miles behind to get here.

Time filters out the scents, the sounds, the touch and the taste of St Kilda on the air, and leaves only the sights (or the memory of some of them, at least), for every one of my fourteen days there was crammed with them. The freedom I was permitted to wander at will and alone meant that I crowded the days and some of the nights with everything all the time, occasionally retreating to my tent and my small portable typewriter to spill out the chaos of St Kilda into the manuscript of my first book. There was no time between the two, and no distance. And now, so much time has intervened, and so much distance; although I have returned often in my mind I have never returned in person. When I consider those first days of that first nature-writing summer (what a place and what astounding good fortune in which to begin), how could I ever improve on all that or add to or embellish it by going back? The simple passage of time, aided by memory's tendency to edit selectively so that only the essential remnants stand forward in any kind of clarity from such a head-on collision with natural forces: all that has distilled down to a single time and place, the indispensable pure gold that sustains one traveller's idea of that time, that place, my own personal St Kilda.

◉ ◉ ◉

The north coast of Hirta is gouged by the sea loch of Loch a' Ghlinne, whose east shore ends in an inconspicuous little headland, Gob na h-Airde. *Aird* is "a promontory (not necessarily high)", according to Malcolm MacLennan's *Gaelic Dictionary*, which is the perfect definition. By the standards of St Kilda headlands it is not necessarily high, nor is it wide and nor is it handsome. In truth, the place looks like nothing at all to write home about, and it would have no place at all in the roll call of St Kilda's natural wonders were it not for what lies under your feet. "Arch", says the map, which is to say "hill" when what you mean is Everest.

There is a bit of a clue in the name of the level ground (and level ground itself is a rare phenomenon on St Kilda) that paves the headland – Leacan an t-Sluic Mhòir – which, like so many Gaelic phrases loses all its intrinsic poetic flourish in the glare of literal translation; but for what it's worth, *leacan* is flagstones, *sluic* is a hole and what you end up with is the "flagstones of the big hole". On these flagstones you can walk across the roof of what is arguably St Kilda's greatest natural wonder without knowing that it's there. To find the "arch" you must go to the edge of a 200-feet cliff and begin to scramble *down*. An easy natural ramp that limpets against the cliff face eases the process. The incredulous stares of razorbills and guillemots at very close quarters do the reverse, for they will imbue your tentative progress with a sense of the absurd. They stand erect like overdressed waiters with their hands behind their backs, waiting, waiting, intimidating, black eyes staring right at you, while all the while they mutter variations on a theme of "aaarrr", with intonations of surprise at the ungainliness of a six-feet-tall intruder; they perch on the outermost edge of the outermost rocks on the outermost edge of the cliff, while the human they mock insists on the innermost, apparently deriving some comfort from the sudden and very welcome presence of a rope handrail fixed to the rock. You cling, you descend, the ocean rises towards you, then the ramp runs out.

Behold the big hole. The very big hole indeed.

But the map did not say "hole" or "cave", it said "arch" and while I am familiar enough with those Hebridean arches that can barely accommodate the passage of a crouching canoeist, it takes a moment to realise that the very big hole is but one end of the "arch". This is "arch" but not as you know it. Welcome to *The Tunnel.*

I once wrote of this place that "memory in such a landscape is not to be trusted", and from this distance in both time and miles, I agree with myself; but here and now, sitting at an oak table in a house in the middle of the country (in other words, just about as far from either North Sea or Atlantic Ocean as you can get in Scotland), memory is all I have to work with. I think I believed then that my nature-writing life would always be like that (I was about halfway through the first week of my thirty-two years in the job), whereas it would never, ever be like that again. For once I got deep enough into The Tunnel to see its far end, where light and ocean poured in from the west, once I came to terms with the fact that the sloping, slippery, rock-fankled floor wanted to pitch me at every step into that unlikeliest of ocean furrows, once I grasped the raw dimensions of the place (imagine an aircraft hangar, the kind where they used to store Lancaster bombers half a dozen at a time), and once I realised that the Atlantic Ocean charged in from both ends and met itself in the middle with the most primitive *noise* I have ever heard before or since (it gives voice to itself in a natural echo chamber the size of a headland)...once I accommodated all that in one brain, one pair of eyes, one pair of ears, the significance of where I was and why I was there and what I was trying to do, and that it was only last Friday I had written the leader column for the *Edinburgh Evening News*, it was then that the otherworldliness of that new life hit home. My response to that overwhelming encounter with nature's company has since become something of a ritual whenever I struggle to cope with the evidence of my own eyes: I sat down. That was when I discovered how wet the rock floor ofThe Tunnel is, and how cold, but I sat anyway, and I stared, and I sat and stared and sat and stared.

And then The Tunnel sang.

Have you ever watched and listened to a wren singing from a perch on, say, a garden rosebush or the back of a park bench, and marvelled not just at the inventiveness and purity of the song but also the sheer volume from a creature the size of a table tennis ball? Surely the inside of a wren is just a hollow, nothing but a soundbox clad in skin and feathers, and a tiny aperture to let the sound out? The St Kilda wren (for it has its own sub-species: *Troglodytes troglodytes hirtensis*, since you asked) is a little chunkier than its mainland kin – "chunkier" being a relative term given that we're talking about wrens – and because it has a consequently chunkier soundbox its song is a little chunkier too. But that assumes that the song is being delivered – as it almost always is – from a conspicuous perch in the open air. The St Kilda wren that sang (and oh, how it sang, and sang and sang and sang) from somewhere deep in that gaunt auditorium, that colossal echo chamber...that wren became choral, became symphonic, for it sounded like a thousand wrens. And here, finally, was the only justification I have ever encountered for the overworked metaphor of nature as cathedral, for it is unarguable at least that The Tunnel is everything in nature that the cathedral is in humankind. It is huge and cool and dim within. And if you carry deep within you the sense of pilgrim in search of sacred ground, try The Tunnel before you give up the search. The light from its vast glassless and unstained windows may blaze beyond its 200-feet-thick walls, but inside it diffuses and darkens. The erratically vaulted roof is of the same mysterious cast that fuses the arts of sculptor and architect. There is also the same sense of atheistic sanctity that appears to dignify ancient stones manipulated into grand gesture wherever you find them. In The Tunnel, a single St Kilda wren was all the anthem nature required.

You can get too much of such a place, or at least I did. I began to feel overpowered, bordering on overburdened. Maybe daylight would help. Then The Tunnel started to moan in a discord of voices, woozy and dizzy-making. There, where the Atlantic meets itself in a vast, head-butting bruise, a dozen grey seals had ridden the surf to caterwaul at the very meeting place of the two Atlantics. They failed, utterly, to drown out the wren and its thousand echoes.

There was still the eastern entrance of the cartographer's "arch" to confront, but by now I had moved so far through The Tunnel that perspectives and sightlines had transformed. Now that colossal yawn doubled as a lop-sided picture frame for Boreray. Boreray never just appears, it seems, but rather it always has to make an entrance, and from The Tunnel it presents one more variation on that theme of sea monster with which I first beheld it at 5a.m. days before, days that felt like years. It would be no surprise at all to hear it roar. But when Borerary materialised in the mouth of The Tunnel, it had finally met its match, the one thunder it could not steal. There are the two most astounding fragments of Scotland's western seaboard, one framed within the other, but that is only possible because Borerary is four miles distant. Grand gesture within grand gesture.

Nothing comes back to me of the walk back through The Tunnel, back up its steeply-sloping floor above the two Atlantics, back out into the daylight of the overworld, back to the rope, the ramp. Were there still guillemots? Still razorbills? What I remember next is sitting on a rock out on the floor of the headland and trying to come to terms with the new knowledge that it is also the roof of The Tunnel, and all *that* was still going on beneath my feet. There was an eerie time of complete thoughtlessness. When it abated (a few minutes... half an hour...an hour?) I was still sitting on the same rock, and I was stupefied by landscape. Eventually, I simply stood and began walking the shore of Loch a' Ghlinne, having resolved nothing, having come to terms with nothing. It would take – and it has taken – the passage of years.

"Shore" is an almost alien concept on St Kilda. Usually, it is another word for clifftop. Other than in Village Bay, you don't walk along shores but along cliff edges. At Loch a' Ghlinne these vary between 100 feet and 600 feet. At its utmost, the western shore culminates in a headland called the Cambir, Hirta's most north-westerly thrust, and when it halts abruptly at one more clifftop the ocean is 650 feet below your feet. Halfway there, where the headland briefly narrows through a wind-bedevilled chicane, the map intones "Settlement" in that bizarre Gothic script it reserves for sites of antiquity. If there is a more un-settlement-like place on the map of my native land, I have never seen it. Yet it was there, on these flanks of Gleann Mor above Loch a'Ghlinne – and not to Village Bay – that the first St Kildans came and settled, planted roots of a kind 4,000 years ago. What were they thinking?

From the summit of the Cambir then, I began to think about where I was, what I had seen, where I had been; about the extraordinary reach of this uneasy coalition of rocks that can still burn a deep unease into the mind of strangers from boats, even after 4,000 years of occupation, even now so long after the final evacuation of the native population in 1930. The rocks still summon the curious, hypnotise them, and yes, stupefy them. From up here, The Tunnel's headland is lowly, the mouth of The Tunnel is nothing more than the mouth of one more cave. Loch a' Ghlinne, from up here, has – apparently – many such mouths. By now, I knew it was a deception, and at that moment, the wind faltered, the afternoon was suddenly warm.

I sat on and on, sun high, the sea slack, and I set myself the task of trying to revisit The Tunnel's introverted underworld in my mind from this high and airy pedestal. It is an old trick. I first began to use it in the Cairngorms, sitting on a high summer's plateau and trying to conjure the midwinter mountain in my mind: ice-bloated grasses, a huge frost-haloed moon, the mountain not summer green but the blue-white shade of ghosts. The method is to fuse those extremes of nature's year in such a singular landscape into a single strand of thought. That way, whenever the landscape is recalled or revisited, I see not the surface of the land in one particular mood, but something bonded and deeper. I find it hard work, and it is perhaps only the particular preserve of the nature writer to go looking for such insight (for there is nothing remotely scientific about it), and it is as elusive as a flat calm on St Kilda. But in the unlikely setting of that aloof fragment of rock – an outpost of an outpost, poised halfway between the sea and the island summit – St Kilda suddenly achieved exactly that, a flat calm, and I urged my mind far back down into that tumultuous airspace, giddy with seabirds, and not a breath of wind. Far below, the sun lit up a small flock of kittiwakes, glittering white against the blackest of blacks where The Tunnel gaped, and then they dived inside.

In my mind I went with them. I was that outrider bird on the outermost edge of the right wing of the flock, the unaccustomed heat on my back. If there is happiness among birds, surely it is to be found in a flock of St Kildan kittiwakes riding the easy rise of thermals on splayed and gliding wings. The flock's entry into The Tunnel was as different from mine as it is possible to imagine. No tentative creep across wet rock in a state of primitive awe at the abnormal nature of the surroundings. Rather, the flock gatecrashed its sanctity with raucous self-confidence, oblivious when the sun on their backs yielded and the world grew dim and walled-in and roofed and cool. They cleaved the heart of its airspace, an undisciplined squadron of wavering, white and black-tipped wings, and the walls and the vaulted roof reverberated to twenty kittiwake throats chanting their own names.

I extricated my train of thought from the edge of the flock and tried to imagine instead how it would have looked if they had flown in while I was inside: *See how luminous the birds are in the gloom. Hear how the individual voices of the flock separate from each other, each responding to the unique acoustics. I can put a voice to an individual bird and every voice spills upwards – as elemental and elusive as spindrift and wind – into the vaults of The Tunnel, drifts back down again and again and again through chains of limitless echoes. "Kitt-ay-wake! Kitt-ay-wake!"*

High on the Cambir, I felt the wind return, saw the ocean stir itself with sudden soft, slow breakers of kittiwake-white. I wondered why the birds would take to The Tunnel at all. I have always associated them with those rocks where the water is whitest, flying eagerly into the exhilaration of storm, relishing the charging air, the booming surf, the ocean's highest, whitest surge. It occurred to me then that they had entered The Tunnel during that uncanny lull, that brief spell of windlessness, of sea idleness. Where, in all St Kilda, would there still be white water, one restless pool of turbulence? Where else but in the heart of The Tunnel, where sea flowed in from each end and the two Atlantics collided.

◉ ◉ ◉

A few months later, at a book event in Fort William, I met one of the last surviving St Kildans. He was in his eighties and he had just been back, just once more.

"We all go back," he said, "just once more, just as long as we are able."

I thought how frail he looked, but how strong within. Like St Kilda.

I talked to him about The Tunnel. I tried to tell him how moved I had been by it. He looked up from his clasped hands on the table then, looked me in the eye, nodded once, gave me a half-smile, and turned his head away. He couldn't speak for a few moments.

◉ ◉ ◉

That, then, was the summer when my nature-writing life began. Before I started writing this, I took that very first book down from my bookshelves and looked at it for the first time in years. The prose reads as if it was written by someone else, and in many ways I am a different person from the one who cast off confidently from the safe haven of the newspaper business on a journey to who-knows-where.

But what surprised me was the poetry. Colin Baxter had asked me to write set-piece passages of poetry or prose-poetry to sit opposite a few individual photographs. Bearing in mind that all I had had written up to that point was journalism, I rather wonder now where I found the nerve to write seven poems. It's not Norman MacCaig or George Mackay Brown, but thinking about it thirty-two years after the event, perhaps these few lines engaged more daringly with the landscape, with their brevity and their sparseness. So in the end I dusted down two of them, "The Old Song", which was at the very end of the book, and "Behind Boreray", chosen because of a remarkable coincidence that occurred while I was writing this chapter. It came in the form of an email from Beauly. It was from a landscape painter called Kirstie Cohen whom I had never met and whose work I had never seen. She was preparing a catalogue for her new exhibition at the Kilmorack Gallery near Beauly. It was called *St Kilda, An Atlantic Journey*, and she wondered if she could print one of my poems beside one of her pictures. The poem was "Behind Boreray".

The Boreray poem was the only one I had written on St Kilda, the only one that came home with me. I was offered a trip round Borerary and the stacks by the skipper of a yacht that had just put into Village Bay. I was helplessly seasick for about three hours, but the seasickness was a small price to pay for the chance to spend time with nature at its utmost, beneath trembling towers of rock and gannets. As soon as I was back on land (the land that still trembled whenever I stood still for hours after), I put the form of words down in a notebook, exactly as they appeared in the book, exactly as they appeared in the catalogue, exactly as they appear at the head of this chapter.

It feels as if the poems have been rescued, allowed out into the daylight again after a long exile in the twilight zone that all books inhabit when they are out of print once and for all.

The Old Song

To have *lived* here,
a hovel on Hirta
for your only hearth
(not nomads of science or soldiery
nor passing prowlers with pens
like me or Dr Johnson),
to bide all your times here
knowing no other's march,
was to look wilderness in the eye
and dare it to deny
your daily bread.
To have *lived* here,
content with all the world
in your embrace, at ease
with all its ways,
then hear compatriots whisper
"Evacuate!" was to feel
the soul's anchor drag,
to know that whatever the voyage,
wherever the final haven,
the journey was done,
the old song sung.

Chapter Two

Forty Years at Eagle Crag

Its survival when all the others went under has made it an emblem of the serenity and the grandeur of the Highlands.

Frank Fraser Darling

The Highlands and Islands of Scotland (Collins, 1947)

IF I WERE A GOLDEN EAGLE I would choose to live here among the southmost mountains of the Highlands, where the Highlands tend more towards serenity than grandeur. As it happens, golden eagles agree with me. Or perhaps it is the other way round. To my certain knowledge they have nested in this glen for fifty years, very probably for much, much longer. My OS map inadvertently confirms that probability with the words "Creag na h-Iolaire", Eagle Crag in the Gaelic language that named so much of the map of the Highlands. Wherever Gaels and eagles have co-existed, there you will find crags, buttresses, rock faces and rock outcrops bearing the same name. And all of these rocks either look like an eagle from an angle familiar to the nearest community (whether still lived-in or long-lost); or eagles are, or once were, accustomed to perch or roost there, or they nest there or thereabouts, or they used to nest there in the heyday of the landscape-namers, until some mishap of nature or some irreparable intrusion of humanity rendered them inhospitable. It's always one or the other and usually the latter. The name adheres regardless. And perhaps, in some such circumstance, a day will dawn when nature effects some haphazard convulsion that will accidentally right an old wrong and a wandering eagle in pursuit of some trait of race memory will pass that way and find the old rock to its liking for all the same old reasons, the *iolaire* will reclaim the *creag* and once again the old rock will live up to its name.

The particular Creag na h-Iolaire that graced one particularly golden golden eagle day in the summer of 2019 has settled back into relatively reliable productivity after a series of singularly unnatural mishaps at the hands of egg thieves. But now that that perversion has more or less gone out of fashion, golden eagles have prospered here, and they continue to prosper. The climate is more benign than it is among mountains further north ("benign"...a relative term, it's still the Scottish Highlands); the eagles' prey is more abundant and diverse and so is the habitat. So this particular Creag na h-Iolaire is a nest site. Like most golden eagles, the pair that inhabits it has three or four nest sites, but there is always a preferred choice, a natural home. The others are fallback options when something goes wrong. Here, the preferred choice is a broad and shaded ledge, overhung, and terminated at one end by a rock wall. It is ideal in many ways, yet in another, it may be about to outlive its usefulness, for the ledge is increasingly imposed upon by a burgeoning bush. When the view of the ledge from far below was completely unimpaired (quite a few years ago now when a watch was kept to thwart the egg thieves), the eyrie itself was visible in good binoculars at quite a distance, a vast, tapering, flat-topped sprawl of twigs and branches and vegetation the shape of a slightly lopsided Mount Fuji. But it is in the nature of eagle eyries to grow more massive every year, and the evolving architecture at work here will eventually run out of shoulder room, run out of headroom, run out of ledge. I can't help wondering, what then? Will the occupants deliberately dismantle what was painstakingly built over so many years and begin again with posterity's preferred option; will they promote one of the fallback sites instead; or will they head out across the glen and cross the watershed to a different mountainside? If they choose the latter course, no one will trouble to change the name on the map.

There had been a story in the newspapers at the beginning of July: two young golden eagles fledged from this eyrie had been fitted with satellite tracking devices. (We do like to put our clumsy fingerprints all over nature, don't we, convinced of our entitlement to know everything about where they go and what they do…the arrogance of our species.) These two had gone missing in Perthshire on the same day, somewhere over a grouse moor estate which has something of a speciality of making eagles disappear. The circumstances were described as "suspicious". Worse, the eagles that vanished had celebrity backers. The tracking project was being monitored by a television presenter and a Member of the Scottish Parliament who had given the eagles Christian names, and the media made a bit of an arse of reporting it, the way they usually do when crimes against nature come to light, and reporting standards only deteriorate further when there are famous names involved. The trouble is that the focus is inevitably on the celebrities rather than the eagles, on their soundbite outrage rather than the wildlife crime itself. And there is an unfortunate and increasing tendency for the celebrities to be filmed or photographed handling the birds during tagging, as if that lent the process some legitimacy; it doesn't, it sets a poor example. A variation of the same phenomenon pervades much of what television wildlife documentaries have become; it is no longer enough to let nature tell its own story in its own languages. Count how many times a celebrity presenter says, "Wow!" or "Stunning" or "Iconic" or (worst of all) "Look at that!" (It's a television set, what else are you going to do with it?) Then weep at production priorities that value thoughtless celebrity English higher than, say, wolf howl, whale song, swan cry.

This Creag na h-Iolaire, with the encroaching bush on the eyrie ledge, is a mighty lump of rock I have known for more than forty years, and thanks to older sages who introduced me to it lang syne, I have known its living-memory history for another decade or two before that. High summer confers on the upper glen something of the air of an alpine meadow. It is more or less ungrazed, "more or less" because there is no such thing as an entirely deer-proof or sheep-proof fence, but the landowner is the Forestry Commission and it knows as much about fencing out the grazing tribes as anyone. Grass is knee-high and sways in the breeze with some of the aplomb (but lacking the grace) of an east-coast clifftop barley field. Flowers flock to the liberated land, and butterflies and moths sway among them with drunken delight. Mountain avens, grass of Parnassus and a clutch of orchid species hunker down among the tussocks and grow brave among the rocks and along the burn and the edge of the woods that dignify all the glen's hillsides. Thyme and thistles and kidney vetch and tormentil work the inexplicable sorcery nature achieves whenever it juxtaposes yellow and purple in random clusters, a gleeful disorder of daring colours that snares my admiration every time, and I wish I could write that way, or draw and paint like that, with that freedom, that daring. But as Paul Cézanne said (in this book's predecessor, *The Nature of Spring*, among other places!):

> *I become more lucid in front of nature...but I cannot attain the intensity which unfolds to my senses. I don't have the magnificent richness of colouration which animates nature...*

If it was beyond Cézanne, I suppose I shouldn't be surprised when it eludes me.

Besides, this was how I assuaged my anguish, how I might ease the burden of grief for two crudely mistreated golden eagles, the ultimate symbol of nature in my country since we exterminated the wolf, and that symbol had been torn down like a despised flag and stamped into the earth by people who would have us believe that they are the guardians of nature. They use the very words. So I came here, to pay my respects to two young golden eagles that were deliberately killed (and if they had been people, we would have said deliberately murdered in cold blood).

A burn of the most potent, flavourful water you will ever

taste bisects the wide floor of the upper glen, having already slithered its way southwards from its source far out on the high moss beyond the headwall. When it reaches the edge it skips and tumbles in short leaps and cataracts, pauses twice in deep and quietening pools, before raising its voice again for the last of the falls above the meadow. Only beyond the meadow does it settle into its stride and its task of becoming a Highland river, then a string of Highland lochs, then a Highland river again, before it slows and unspools into the long eastward haul as a Lowland river towards the distant Forth. But back up on the moss, where it rises out of a land of snipe and adders and golden plovers, a second water rises not a hundred yards to the north, but it will cruise different lands, for the moss sheds south to the Forth and north to the Tay. On its north-making way, and before it heads east for Loch Tay and my far-off hometown of Dundee, that second burn cruises past a small and unhappy cottage where I once passed a few years in pursuit of a changed life.

I had climbed to the headwall from the upper glen and headed out to that double-edged watershed, for I never became immune to the impact of its miles-wide view of northern mountain walls, a view that became a fixture of the cottage years. My unease about the two lost eagles had rather got under my skin and it was still troubling me as I breasted the headwall and the mountainous north suddenly reared up and smothered the horizon. At that moment I was taken unawares by a sudden sense of the freight of lost opportunity that those years and that view had once promised. I thought that I had been healed long ago, but now my mood appeared to have found its long-lost kindred spirit in those mountains, for they hold several golden eagle territories. Three that I know of are visible from that same spot between the rising of the two burns, and a fourth, for the moment, lay behind my back with Creag na h-Iolaire for its hearthstone and its heartbeat.

So I sat and grew still and looked at the thing that had just troubled me, and turned it over in my heart and my head to see what I might make of it. And now that a little more time has passed and I have added it to the fifteen years or so since the cottage and I parted company, it seems to me that the trouble stems from the fact that what had helped me most through those troubled years was working with the eagles of those glens, walking their mountain heartlands in every season, learning a little of their lives and a few of their subtleties and secrets; not by fitting satellite trackers to them but by increasing my chances of encountering them – and their fellow travellers – by going out there into their territories again and again and again, piecing the inevitably brief encounters together to fashion them into a kind of mosaic of understanding. And from time to time, the eagles from two of those territories turned up on the slopes above the cottage, so that I could watch them from the back garden or a rock by the burn not a hundred yards away – that burn which flowed south from the watershed on the high moss – and those encounters had felt at the time like a kind of reward, and I held them particularly close; I still hold the memory of them close, too.

That summer day of July 2019, then, somewhere between the rising of the burn of the cottage and the rising of the burn of the glen of Creag na h-Iolaire, I reasoned that surely enough time had elapsed to allow me to recognise the symptoms of what had lain buried, then reminded myself about what had healed me, and that I was back here now to seek the reassurance of eagles, the fellowship of eagle landscapes, and to lean on them that I might step closer towards them again, as I had done so often in the intervening years. Sometimes, when you live that close to nature, you become nature yourself, or at least nature's society becomes your first loyalty, and loyalty to your fellow mortals is a diminished force. So I edged south again to the rim of the headwall, found my favourite rock there where I have sat so often over so many years.

It was here, in these hills in general, but at this very rock in particular, that I learned from the eagles how they worked their own idea of territory, how intimacy with that territory in all its seasons and weathers served their purposes and enhanced the apparently limitless repertoire of the intricacies of flight, so that no creature knew that land and its skies better, nor worked with more confidence the spaces between mountain walls. I was so impressed by them that I adopted my own idea of a writing territory, and in my own un-eagle-like way, no single idea has better served my nature writing cause. I am not indifferent to the debt I feel to golden eagles, nor to the landscape that spills from that rock on that watershed to every compass point.

◉ ◉ ◉

When you walk through the upper glen, Creag na h-Iolaire looms, drags your eye towards its silhouetted eminence, apparently the summit of the visible land. The view from the watershed dispels the illusion. The buttress is nothing but a geological blip, a little less than 1,500 feet up a 3,000-feet mountainside. The watershed also elevates you into the airspace of eagles, and the sensation of standing at the very heart of the eagles' territory is irresistible, for there are miles and miles of it garbed in every shape and shade of mountain country, and there are all those skies. It was here that I finally learned the true definition of the word "spreadeagle".

In early summer, when the eaglet is readying itself for first flight and the buttress is the focal point of the resident pair's workaday lives (although it pushes the boundaries of imagination for this watcher on the watershed to dismiss any trait of golden eagle summer as "workaday"), the adult golden eagles often fly across the glen or fly the length of the glen, and pass across the face of the buttress again and again, sometimes in slow circles, sometimes in short dives and climbs. There is nothing quite like a blaze of summer sunlight to enliven such flights. Then, when the bird homes in on the buttress and enters its shadow, it is as if the glen's brightest light has just been smoored. The bird won't land, it wants to lure the eaglet out from the sanctuary of the shaded rock out into the bright unknown of the sunlit air, to convert the tentative wing-threshing vertical bounces on the eyrie ledge into the wonder of golden eagle flight, the most glorious, the most masterful flight of all. It is a breathless moment for all of us...eagle, eaglet, watershed watcher. I saw it only once, that first flight, and if it was something of an anti-climax (it was steeply and slightly chaotically downhill and ended with a graceless and gruesome landing on steep hillside grass). About 200 unconvincing yards. But it was a start, it was flight, and for all that it is simply ritual in the eagle's year, for me it was indelible, unforgettable, the more so because it was watched from above, as an eagle might watch it. Yet given the eaglet's complete absence of spectacle, it would be almost impossible to impress on any casual hillwalker or Munro-bagging obsessive anything at all of its significance.

◉ ◉ ◉

She came in low and slow and out of the north-west. I saw her approach from a quarter of a mile away and from behind, over my right shoulder, because I had half turned in response to an unexplained summons, the source of which was not yet clear to me. She was not hunting. At least, I don't think so. It is not always easy to tell when she flies slowly. She often hunts within a yard or two of the ground, a phenomenon of flight I don't pretend to understand, for she flies so slowly and at times her wingtips almost touch the wingtips of her shadow and that seems to me to contravene every principle of flight, which merely shows how much a golden eagle knows about the principles of flight that I don't. But usually when she hunts, she flies slower than this, and her head is down and the wings are held wide. It is tempting to say her wings are stiff, but I have no idea whether she holds them stiffly or whether their stillness is utterly relaxed. But when she came in low and from the north-west, she was not flying with that intensity. Instead the flight appeared lazy, wings beating to a leisurely pulse.

The sun was full on her, and of course I looked for her shadow – I have always loved to watch it ripple over rock and grass and just behind her or just ahead of her, just uphill or just downhill. But on that early July afternoon it was almost directly beneath her, the sun a high glare that lit her fabulously from stem to stern and wingtip to wingtip, with that delicious "golden" blaze at her nape. The shadow was so dark it looked like black silk as it crossed the watershed land. Her tail, half open, rocked gently, the least movement you can imagine while still being certain that it was actually moving, millimetrically precise.

She was not more than 200 yards away when she lifted suddenly. In good binoculars and so brilliantly lit, it was a godlike gesture. At once there was distance between her and her shadow. I thought perhaps she was anticipating the changed conditions beyond the watershed, the effect of thermals beyond the headwall of the glen. But from 100 feet up she flipped onto to one wingtip like a barn owl and charged the ground, only to whip up again in a tight curve, unbeating wings held in a shallow up-curve, and I saw the tall grass beneath her rock in her wake. From the top of her climb she hauled in her wings to half-closed and surged forward at speed. In that guise she crossed the headwall and her shadow vanished into thin air – thin, warm, thermal-fuelled air. She cleaved through it as an otter might cleave slack water.

As I watched her dwindle into the middle distance of the glen's highest airspace and as she eased down into a shallow glide towards the eyrie buttress I lost her against the dark hillside and its confusion of scree and rock trees. At once there was the sound of a scuffle behind me, somewhere out in the boggy watershed terrain.

Turn slowly.

Remember you are trying to be a part of the landscape. Somewhere about the point at which the eagle had climbed and dived down and climbed again, there was a mountain hare, sitting tall and with its ears tall too, staring after her. Almost at once, a second smaller hare rose to sit beside the first, just a few inches away.

Questions:

How long were the hares hunkered there?

How long had they been aware of the eagle's presence on the hillside before I was? They have virtually 360-degree vision thanks to protuberant eyes set high on the sides of their head, so I am guessing they knew long before I did.

Did they recognise the absence of lethal intent in the eagle's approach? Are they so attuned to every twitch of their environment that they can make such life-or-death decisions and opt for stillness rather than a chancy dash over open ground before the rim of the headwall offered steep, tree-and-rock-strewn cover that would surely thwart the eagle, but only if they got there first? It's a big "if": a flat-out mountain hare can build up speed to forty miles per hour, a golden eagle in a shallow dive to more than twice that. It's a tricky call, trying to out-bluff a golden eagle.

And one more question:

Was the golden eagle's dive to where they lay in the long, pale grass nothing more than mischief-making? If not, why did she pull out no more a yard above them? Is there something in the eagle's prey drive which, like the wolf's, is only stimulated if the prey species is on the move? Did the hares know that? Surely the answer to that is "yes", given how long the two have evolved together and the sophisticated nature of the relationship between the two. And if it's true, and if the eagle *was* hunting then wouldn't their best defence always be stillness? The tricky part arises if the eagle's response is not reliable. Then there is literally no hiding place for a mountain hare that chooses to live within a mile of a Creag na h-Iolaire.

These thoughts tumbled through my mind as I turned back to find the eagle, but she would not be found. So I turned again to the hares but they, too, were gone. In half a minute of looking the wrong way – twice – I lost both eagle and hares. I have been told many times that the amount of wildlife I see is impressive; I have been described many times as "a wildlife expert" (I deny the charge). Here, if I ever needed to be reminded (I don't) was all the evidence anyone could wish for of my limitations in such acutely attuned company as this.

So I did human things. I sat, poured coffee, munched chocolate, stared out at the mountain land so familiar to me now after forty years and considered what had just happened. I have good eyesight but I had not seen the hares arrive in the hour I had been sitting watching the mountains beyond the watershed. Perhaps the hares were already there, resting up in the middle of the day and it was that long stillness that worked in their favour. In which case it is equally true that my long stillness had worked in mine. Neither eagle nor hares had reacted to my presence. The eagle, when it had finally crossed the headwall, did so no more than 100 yards away from me, and without a second glance.

I sat on for a while, just because it is a good place to sit, but nothing more stirred.

I picked my way down the headwall by the burn, slaked my thirst and cooled my head and neck in a pool between two waterfalls, paused where the path and the burn part company for a while at the foot of the headwall, scanned every corner of sky and glen and middle distance, of crag and buttress and woodland. No eagle showed. There was a glimpse of fox crossing a patch of sunlight on the far side, and it struck me that for all the toll the eagles take of cubs, adult foxes have denned in the same part of the glen within clear sight of Creag na h-Iolaire for decades, and possibly centuries, and still raise enough cubs to justify their loyalty to the site. Foxes belong here. I admire them, their wisdom, their tenacity, the way they always find the means to outwit our worst efforts to obliterate them. Every time I see one I feel like cheering. I always wave.

Just where the upper glen begins, I turned to look back. How different the glen looks after a few hours: the deeper texture of the sunlight, the advance of shadow through the afternoon. The birches in the highest part of the glen looked almost smoky. I lingered over them in the glasses, especially the darkest corner where they lean against a great slash in the bare rock of the mountain. A flicker of pale colour in there: what? Elbows on a rock to steady the glasses, for it was a long way off. There, again, the same movement. There was a young golden eagle on a bare branch, and as it moved with the wind up there, it opened its wings to hold steady, and it was that, that revealing of white wing patches and the way they disappeared when the wings folded again, just an abstract moving shapelessness but almost bright white in the shadow of the crag: confirmation, reaffirmation, gratitude for the watcher. The mouthed silent "thank you" was directed at nothing, everything, at nature, and I waved again, to all of it – the watershed, the northern mountains, the mountain hares, the headwall, the woodland, the foxes, the flowers. And to the eagles of Creag na h-Iolaire for all that I owe them, that debt of gratitude accumulated over four decades.

⦿ ⦿ ⦿

A slow walk-out and a cooling beer later, I considered the nature of my response to the disappearance of the two young eagles over that Perthshire estate with its grim grouse-as-be-all-and-end-all style of management – where the end justifies the means, and the end is making money by killing grouse and selling them to people with more money than sense, and the means includes two dead eagles that disappeared in a cloud of celebrity indignation. I don't deny the celebrities their right to feel indignant. So do I. But it doesn't help. Despite the satellite tags, despite the celebrity sponsorship, despite the laws of Scotland that afford golden eagles the highest level of protection from the worst excesses of grouse moor and deer forest estates, the eagles died anyway.

Eagles flourish where they are left alone. Landowners and their gamekeepers should leave them alone. It is stating the obvious to suggest that penalties should be much tougher and that resources to fight wildlife crime should be much greater. And celebrities should certainly leave them alone, but they can only interfere in the first place if professional conservation permits it. And while it is not a fashionable opinion, I believe that professional conservation should leave the eagles alone, too. We don't need to know their every movement. We don't need to handle them. Nor do we need to remove young birds from Highland eyries and take them to the Borders in the faint hope that human intervention will boost the unsustainably small south of Scotland population of two or three pairs most years. But this is happening now in the South of Scotland Golden Eagle Project. Highland eyries with two chicks supply the birds, one is taken from each of three eyries at between five and eight weeks old, and moved to specially built enclosures in a quiet part of the Borders where they are artificially fed until they have fledged. Then they are released in the Border hills.

Problems:

One – The birds are deprived of being taught to fly and hunt by their parents, the natural birthright of every wild eagle.

Two – The Highlands and Islands of Scotland constitute the golden eagle's stronghold. Every shred of common sense tells you that the priority of sound conservation thinking is to protect the stronghold. There are any number of empty territories for a growing number of golden eagles within the stronghold. And nature tells you the same thing: if golden eagles wanted to colonise the Borders, they would choose to do so. It is no great distance from Arran or Kintyre, for example, to Galloway and the western Borders. But the eagles choose not to, and they have consistently chosen not to for more than a hundred years.

Three – the reasons they choose not to are self-evident: lack of suitable habitat, lack of suitable food, and too much disturbance. You can add into that unpalatable mix the superabundance of wind farms. On a clear day from a hill-top in the release area, looking through 360 degrees, you can count 700 turbines.

Four – what makes us think that golden eagles sourced from the Highlands and released in the Borders will stay in the Borders, rather than, say, flying back to the area where they were born?

The thing is this: we are not better at deciding what is good for golden eagles than nature is. Our obsession with ringing, wing-tagging and satellite-tagging does not produce better conservation. Rather, it produces alarming levels of intrusion for both young and adult birds at the nest when the devices are fitted. It also promotes the increasingly widespread attitude that our interference is good for nature. But our track record of managing wildlife is not impressive, and of managing land it is simply wretched.

Not for the first time, permit me to quote American nature writer, artist and wetland specialist David M. Carroll, from his book *Swampwalker's Journal* (Houghton Mifflin, 1999):

> *The term "wildlife management", often used in the environmental polemics of the day in reference to human manipulations, is an oxymoron. We should have learned long ago to simply leave the proper natural space, to respectfully withdraw and let wildlife manage wildlife.*

As I write this, there is a golden eagle dying very slowly and in agony somewhere near Crathie in Aberdeenshire, or it may already be dead. It was photographed in flight by a visitor, and it had an illegal trap on one of its legs. It is not hard to reconstruct the sequence of events. The trap is set out on a remote corner of a Deeside grouse moor. The trap is baited, probably a dead pheasant or a piece of deer. The eagle lands beside the bait, the trap springs, but instead of killing the bird it closes on its leg. The eagle's flight is now clumsy, slow, off-balance, and it will endure immense pain until it dies. One more eagle wearing another artefact of human technology. Granted, the purpose is grotesquely at odds to a tracking device, but it's all wildlife management, designed to achieve an outcome predetermined by people, not nature. There is not that much difference in the arrogance at work: we know best.

Let wildlife manage wildlife. It is the simplest, and the purest of ideals. And it absolutely terrifies us.

Creag na h-Iolaire, remember the name. It means Eagle Crag wherever you find it. It's not my rock, not yours, nor the landowner's nor his keeper's, not professional conservation's, not a TV celebrity's. It's the eagle's rock. If you are ever near one, respectfully withdraw, leave the proper natural space. Let wildlife manage wildlife.

Chapter Three

You Have Not Seen Her With My Eyes

THE RUSH. THE HOARSE ROAR. The rush is dead ahead. There is a scar like a furrow on the surface of the water. No, just beneath the surface. But, unlike a scar or a furrow, it advances – in a rush. The hoarse roar is slightly stage right. Ears try to drag eyes that way. The eyes think about it but not for long. This, the rush, this whatever it is that rushes and scars and furrows the surface from just beneath it, this is better. The eyes will not be distracted. The sheer speed, but also the control...they insist on it.

The eyes see silver. Not breaking the surface, not quite, not yet, although at every moment the tension quivers and surely the surface will burst apart now.

The eyes think, "Torpedo". Then they start to anticipate "Explosion". But not yet.

Instead, the speed increases. But the control remains precise. It all looks impossible. At least, it does to me. But I don't fish.

I don't fish and all I know of torpedoes came from war films a long time ago. And that rush – dead ahead, not quite breaking the surface, that silver, that shapeliness – that is the prototype. Someone with a scientist's mind stood on a riverbank like this one and saw this and thought, "Torpedo" and then "Explosion". I don't have a scientist's mind any more than I fish, which is not at all, ever. But my father was a soldier for ten years before and then all the way through the Second World War, and although he never talked about it (and there was a lot to talk about, what with Dunkirk, El Alamein, his admiration of Montgomery, Sicily, Berlin at the end; and I have his Victory Parade programme and photographs of him with a half-track truck he had driven in the desert), he relived something of it with his family: we were taken to see any number of war films, and as my maternal grandfather managed three local cinemas in the Lochee area of Dundee, it was easy. So there were films of the Army's war, the RAF's war and the Navy's war, and every now and then there would be a submarine in the film and then there would be torpedoes. So to that extent (and no further), I knew what a torpedo looked like just below the surface – the rush, the shape, the speed, the control, the explosion. And this that I could see now with enthralled eyes, this was the prototype. Dad never dreamed that a legacy of his soldiering would bear such strange fruit.

And then my eyes began to think, "When? When?" The explosion. Any second, the explosion. But still it came on, the rush, and the water caved in before it as its momentum cleaved the surface from beneath. Still it picked up speed, yet its torpedo shapeliness was pitted against the spate-fuelled charge of the river, against the press of tons of fast water still turbulent from the upstream chaos that generated the hoarse roar. I had a sudden intuition: an unlikely symbol, a new metaphor (new to me, at least, though maybe it's in the back catalogue of, say, Neil Gunn!) – the golden eagle of the river, because am I not forever writing how the golden eagle achieves control and speed when it fires itself upwind? This is the golden eagle's way, too. Golden eagle, Atlantic salmon; who would have thought they would summon the same techniques?

Then it turned. I never saw a torpedo turn before. No World War II movie torpedo ever turned, and those films were the beginning and end of my knowledge of torpedoes until this, this cleaver of the upstream summer river, the morning after days of rain ended and sunlight poured onto the river and the water was black except where it was dazzling white, the dazzling white in the river that was the source of the hoarse roar. So the Atlantic salmon opened a long curve that cut into the quieter waters of the near bank, and by that curve the hoarse roar came into play, for it was the roaring, dazzling, white water that had to be overcome, the roaring water that challenged the salmon's very right to exist, to return far into the upper river that spawned it. Unlike the torpedo, the salmon's explosion was designed to overcome the hoarse roar, not to sink it but to silence it, by leaving it behind. The curve widened then tightened. The furrow, that scar that was the salmon's torpedo spoor, lengthened and burst apart more of the surface as the speed increased, then the curve was done and the rush straightened and headed for the very heart of the maelstrom, that hoarse roar that was nothing less than the massed voices, pipes and drums of ten waterfalls stretched across the breadth of the river, and these pooled their thunderstorm resources in such an upheaval of white water and sunlight and sound.

Then the salmon dived and was gone.

◉ ◉ ◉

There is a hole in the river, an ice-gouged hollow that has since become a broken, rock-walled sink. Just as the eagle glen has a headwall (the analogies between the two tribes do seem to line up to be inspected), so the salmon's river presents a landmark headwall that rears above the known landscape, and it is that which the salmon must negotiate. The river, even in this mood of bloated tumult, could not overwhelm the wall, but rather divided around it to left and right. So, crammed between the broken ends of the wall and its own alder-lined banks, it tumbled sideways into the sink by way of ten separate falls that collided and co-mingled in joyous reunion moments later in the deepest depths of the sink beneath the headwall, and the colour of that reunited river was sunlit snowdrifts, so that the whole spectacle almost dared you to look it in the eye.

The river's problem now was how to get out again. Almost all of it was forced (under what kind of forces and at what speed?) into a narrow channel of imponderable depth where the downstream-curved wall of the sink had long-since burst apart. That channel, just a few yards wide, was suddenly commanded to bear the burden of almost the entire river which was twenty times as wide before it hit the headwall, and the resultant volume, speed and power would reduce you and me to ruins, broken in many pieces. Yet it was this channel the salmon must swim through – upstream – in order to effect the explosion from the innermost depths of the sink that will carry it to quiet waters above. That explosion was the only thing that could silence the hoarse roar.

Deep beneath the ferment of the sink, the Atlantic salmon powered forward with utter certainty. He knows, too, that even here, the deeper he goes the less frenetic the water. How deep is that hole? I have no idea. The salmon knows, of course, just as it knows that the line it held through the channel is also the line that it must hold across the sink, because it will angle him directly towards the widest of the falls, the one immediately to the right of the headwall, the one nature has chosen for this singular fish. I am guessing this part because I can't see it, because nothing else can explain what happened next. I am guessing, too, that the salmon's self-confidence in its ability to hold to that line without the slightest deviation is justified; that although the water pressure is at its utmost down there, it is vertical rather than upstream or downstream, and the salmon has the power and shape and that faith that moves mountains and silences river roars, and these fuse into unstoppable energy. And from the bottom of the sink it builds a new momentum and tilts upwards until it is all-but-vertical then bursts the water apart and heads for the daylight. It knows (it remembers!) there is a shelf two-thirds of the way up the fall to the right of the headwall, and that is its only chance. The explosion when it comes is calculated to thrust him up and out of the throat of the roaring water to reach the ledge where he might just hold still long enough to gather new momentum and go again.

That, at least, is how it looked to me, but I don't fish. The salmon's mind is mysterious to me. Its destination seems never to be in doubt, nor is its ability to find its way. Consider this particular salmon up to the moment that it dived from my sight, deep into the channel, for its journey had begun the moment it decided it was time to go home, and thousands of miles out in the Atlantic from here. In due course, which covers innumerable hazards both natural and man-made – contrary currents, warming seas, predators, ships, nets, plastic in all its grotesque forms – it must find the island-strewn west coast of Scotland, then from out of that strewment it must select Mull. Having found it, it must decide how to bypass it, the north coast through the Sound of Mull or the south coast and the Firth of Lorn. A good map or a natural vantage point like the high ground of Lismore would seem to suggest that the Firth of Lorn is the obvious choice because it leads more naturally to what follows. But is there anything at all obvious about the return of the Atlantic salmon from mid-ocean to the fragment of moving water where it was born? I have thought about this quite hard, and I can think of nothing about the process that I can point to and say, there, that bit's obvious. Except that only by arriving at that same fragment of water far up some tributary of the upper River Orchy can that fish perform the single task for which it is hard-wired by nature: to be part of the process of producing the next generation of Atlantic salmon. It either dies in the attempt, or it succeeds precisely because it is an Atlantic salmon, and the demands of its whole life's journey are dedicated to this here and now. Nature asks a lot of this fish.

So say it chooses the south coast option then swings north up Mull's east coast. Why would it then resist the wide-open jaws and the tidal pull of the Lynn of Lorn and Loch Linnhe beyond? Why instead seek out the narrows of

Connel, the Falls of Lora beyond and the first sinuous miles of Loch Etive? And then, with all the wild world at its disposal, how does it identify the unremarkable entrance into the skinny little River Awe?

The River Awe is a minnow of a river in the kind of watercourses to which this salmon is thirled. It is about five miles long and serves only to allow the waters of the mighty Loch Awe to make their way to Loch Etive and the ocean. So now, and having survived the man-made indignity of the barrage at the Pass of Brander (as grim a place as the Highlands ever devised, deep in the shadow of Cruachan), how tempted do you suppose the salmon is to turn south and explore the waters of the longest freshwater loch in the land, the twenty-five miles of Loch Awe, one of the natural jewels of mainland Argyll? Of course, the salmon knows better, because it knows where it is going and that, literally, it has no choice. So, in pursuit of its allotted task, it crosses the very top of the loch from west to east, then leaves it again by its north-east corner. It has found the River Orchy. There are a handful of relatively low-lying, relatively douce miles of Strath Orchy before the salmon finally veers north-east into the instantly turbulent and pulse-quickening river of Glen Orchy. From here it will have alders for company every mile of the glen, for they line and shade both banks in single file with more than a hint of a guard of honour about them. It is nothing less than a prodigal Atlantic salmon deserves on the last leg of the final homecoming. Welcome home to Glen Orchy of the Falls!

In the midst of the ten falls then, a torpedo is on the move. The prelude is only guessable, and anyone's guess is as good as mine. But, at last, the fish is there and all but vertical, canted a few degrees forward, the tail winding up the momentum even as it finally explodes from water into sunlight and spray, and oh how it glitters in those few seconds of something like flight. And then it seems to reach forward, its bodily attitude lowers towards the horizontal and it strains its every phenomenal resource into a last thrust of forward motion and the waterfall seems to rise to meet it and it hits the wide ledge squarely in the middle and, with who-knows-what manner of gymnastics, it holds on where it lands, and it does not slide back.

The lull is breathless. The waterfall might have the good grace to hold its fire for a few heartbeats in recognition of the leap, the landing, the tenacity to cling. But good grace is not in the nature of waterfalls. It batters on down, and it roars on and on. But now, suddenly, in the world of the Atlantic salmon, the roaring and boiling and water-furies of the sink are behind it. From here, from the ledge two-thirds of the way to sanctuary, it knows the way. In its own time, it will find the second momentous thrust, the lesser explosion, and it will cruise upstream and away from the tug of the last great obstacle in its path. There are other waterfalls on the Orchy, and the power of the river in this mood is fierce along its entire length. But this is what the salmon is good at, and the golden-eagle-like control that comes with swimming upstream reasserts itself. A deep pool lies in an eddy of the bank. Alders cast their shadows along every yard of it. Perhaps there is where the salmon slides into the gloom. Draws breath. Rests.

⊙ ⊙ ⊙

"The least sympathetic of humans must admit a touch of awe," wrote Marion Campbell in her exquisitely pitched portrayal of her native land, *Argyll – The Enduring Heartland* (Turnstone, 1977). I freely admit to that touch of awe.

"Even 'the wisest fool in Christendom', King James VI and I, could write of the 'salmon-like instinct to see the place of his birth and breeding' in planning his one journey northward after his ascension to the English throne."

James, having thus ascended, moved his entire court from Edinburgh to London in 1603 and only returned once before his death in 1625. The surprise is that he invoked the salmon, that its way of life and compulsion to return to its native airt was well understood even then, compared to, say, the wilful lies and ignorance that attended the lifestyle of the wolf.

Marion Campbell lived at Kilberry on the Knapdale shore, as did her family for 400 years. She was also my friend. My conversations and correspondence with her over the last decade of her life offered me rare insights into Argyll's – and Scotland's – past, present and future that would have been quite out of reach to me otherwise. She had wanted to call her book *My Argyll*, which is nothing less than the Argyll of 5,000 years of its story, and the undateable years of legend. She defended the notion in her introduction with a Gaelic phrase from "our poet" Donnchaidh Ban MacIntyre: *Chan fhaca tusa i leis na suilean agamsa* – "You have not seen her with my eyes". That was her so-generous gift to me: that she let me see something of Argyll with her eyes. "Heartland" as she referred to it, is simply one of my very favourite books. I have been reading it for more than forty years. It is among the most elegant books that anyone ever wrote about any aspect of Scotland.

She was writing about Glen Orchy when she invoked James VI and the salmon, and in a curious way that at least has echoes of my own tendency to home in on Glen Orchy, its river, its falls, its alders, its pinewood remnants, its otters and golden eagles, and one in particular of its mountains. I have been doing that for almost fifty years. Marion died in her eightieth summer; 2019 was her centenary year, 2020 would have been her one hundredth summer. It was a good moment to go back to Glen Orchy, to make a pause of gratitude there, to watch a salmon confront a meeting of waterfalls (for these are always portentous places).

In truth, the Marion Campbells of this world are beyond time. She was writer, poet, historian, archaeologist, and a living, breathing fragment of the land where she was born, lived and died; her family lived there in an unbroken line for four centuries, and they live there still. And the themes of *Heartland* endure. The River Orchy still "roars with the voice of an angry ghost in a long-severed Celtic head". The Atlantic salmon still homes in on Etive, Awe, Orchy. The Atlantic itself still laves Kilberry's shore, tide after tide. Tides are a recurring symbol. She planned an autobiography, called it *Tidemarks,* wrote one page – about the tide beyond her window – and abandoned it. But you could argue that *Heartland* is her autobiography; or, at least, a lyrically realised self-portrait for her voice is everywhere in its pages. Just as the river *is* Glen Orchy, so

Marion Campbell *is* Argyll. She was Marion Campbell *of* Kilberry: she was of that place.

"Yesterday Was Summer" is her title for a chapter on childhood, but no sooner has it established the timescale of her lifetime than it extends its reach, because "...my grandfather, born in 1844, kept a diary from boyhood to death... he knew a man brought up by an old woman who remembered 1745". *Heartland* set sail with the first currach and the hunter-gatherers, but its final chapter is called "Tomorrow" and is an unambiguous endorsement for Scottish independence. And "...the long sigh of tide-turn is heard already".

The consuming project of that sphere of her life where historian and writer united in common purpose was a biography of Alexander III. It took her forty years, on and off, before *Alexander III – King of Scots* was published in October 1999, appropriately by The House of Lochar on the Argyll island of Colonsay. She died the following June, her work done. I miss her still.

◉ ◉ ◉

If you are an Atlantic salmon in the River Orchy, mostly all you will see of the world beyond the banks is alder trees. Yet if you are walking or cycling by the river, it may be that you don't notice the alders at all. They line both banks in single file, leaving only enough space between trees to accommodate each other's unimpressive canopies. Behind them, the aspens, the oaks, the birches, the Scots pines, the willows and the hollies compete for the level ground of the flood plain, and crowd down towards the alders; and behind these, millions of spruces and larches clothe hillsides most of the way to mountain tops. The forestry industry has been none too gentle with Glen Orchy. But set against so much swarming summer greenery, the alders are so inconspicuous they practically vanish. Most of the time, wherever they grow in any kind of profusion, vanishing is what they do best.

But what they lack in spectacle they compensate for in tenacity. They hold all the other trees back and for two reasons. One is that they are thirled utterly to water, to damp places, wet places, waterlogged places, and yes, to the outermost edges of riverbanks. The other is that they have evolved a technique to safeguard their right to the role of guard of honour for the epic marchpast of the Atlantic salmon. The tree produces two kinds of catkins: long dangly ones, which are male (sounds about right), and little cones, which are female. The cones hold the seeds and the tree drops them in its own roots. The result is that either they seed there and guarantee the next generation of alders on the parent tree's stance, or the river washes them off and carries them to a new landfall further downstream. So the continuity of alders is fed from the water, and no other tree species gets closer to the riverbank than that. The alders of Glen Orchy, then, line the riverbank in perpetuity. Wouldn't it be nice to know when they first moved in, for the alder is a pioneer species? It would have been at the head of the queue when the last ice age ended around 10,000 years ago, and given that Glen Orchy would have been an early landscape feature as the glaciers from the icecap that was Rannoch Moor began to reconfigure the central and west Highlands into something we might recognise today, the alders of today's river may well be the direct descendants of some of the earliest trees to re-establish in the wake of the ice. So surely they are worth a second glance. There is one in particular that I have in mind.

For all the inherent modesty with which nature has imbued alders like these, the most beautiful tree in all of Glen Orchy is an alder, and a small tree at that, even by riverbank alder standards. It is so edge-of-the-bank in its stance that, with the river in this extravagant mood so that it roars a tenfold waterfall at an Atlantic salmon, the alder wades. Here is where the roots splay neatly then steepen down over rock into the river's peaty embrace. But many alders wade. What makes this one beautiful is its compact and balletic pose, which its water's-edge stance only enhances. The trunk rises straight for only about six feet, at which point it bends at forty-five degrees and reaches up and out over the shallowest water. But only four feet up from the roots, the trunk also throws a limb out towards the water and almost at right angles, and that limb begins at once to bend upwards. The result is a sinuous profile formed by the two limbs, the first staying close to the shore, the second reaching for the mainstream. The alchemy they achieve derives from the shapes of the two spaces enclosed by the two limbs as they curve towards each other and then apart from each other. One is a perfect oval, the other an elongated triangle. Change the angle of your viewpoint slowly and watch the shapes of the spaces change too, widening or narrowing, shortening or lengthening. The hypnotic effect is enhanced by the fact that the spaces enclose nothing but brightly sunlit and vigorously mobile river water. A gleeful wind shivers the leaves and lesser branches of the tree, yet their modest spread has a controlling effect on all movement. So the whole tree is anchored by the trunk but the limbs and the branches and leaves shift subtly against the background of the river's liveliness, and that liveliness is there in the spaces within the tree. I could watch this for hours. With a guitar in my hands I might have coaxed the tree's dance to make music. (I must go back with a guitar sometime and see what I can make of it.) As it was, all I had in my hand was a pen and a notebook to write it down.

I had brought my bike. When I rode on, relishing the midsummer quiet of the glen's delectable single track road, I lingered over the sense of the tree, its swaying image gently seducing the memory as I tried to keep it bright in my mind, the zest of it, "the touch of the daylight on the dream". If you think that an extravagant response to a smallish alder tree on the bank of an Argyll river, you have not seen her with my eyes.

Chapter Four

She Is of the Woods and I Am Not

THE PLACE OF TREES in the landscape is never far from my mind: the dearth of them in Highland and Island Scotland; the devastation inflicted by hill sheep farm, deer forest and grouse moor, which denies trees their rightful place in the landscape; the absence of anything like a natural treeline (restore that and we begin to heal the land at a stroke). I revere, I cherish well-wooded places. I stow away among them often.

Sometimes, especially in summer, I feel the need to seek out cool places beneath old trees – oaks or Scots pines, for preference. And because it is summer and because I live in Scotland, and because summer and Scotland and trees are the component parts of an equation that equals midges, I like my cool places beneath old trees to be open spaces, and high rather than low-lying, and near enough to the edge of the wood to permit access for the wind, the better to keep me cool and move the midges along.

Here is one such open space, in a pine-and-oak wood that reaches a thousand feet uphill on both sides of a mountain river to a trinity of waterfalls, so water voices carry to me on an easy wind, and that adds to the allure. But right now, I just want the good Highland summer green that dresses the space beneath the trees. I know from many years of practice that this here-and-now with my back to an old oak trunk, sitting as still as I know how with the wind in my face…this open invitation to nature to drop in and spend the time of day with me is as likely to bear fruit as absolutely anything else I can think of.

The Wind in My Face

The wind at my back,
my scent and sound
blown before me
and nature a locked door.
The wind in my face,
my scent and sound
a shredded wake
and nature an open gate.
Enter, ask for owl eyes,
tree stillness,
now make what you can
of her secrets.

Then the roe deer. The doe. She walked from tree shadows, where dusk had begun to gather, into the one patch of that small clearing where the last of the sunlight caught her, and there she stopped and there she stood and there she glowed. But the reddening sun was only on her head and neck and spine and the top half of her left flank. The rest of her was dark. She presented an almost eerie illusion, as if the top half of a deer was swimming through the trees and the woodland understorey. When she stopped, she floated.

How long before she would see me?

She was still then, apart from her ears, and these flickered with a restlessness that belied her outward calm. The ears looked too big for the size of her head and they were pale grey, almost white, when they turned towards me, and trimmed with a sharp lace of black. With these she tested every airt of the woodland and far beyond. But I knew the wind worked in my favour. So did my stillness and my woodland-shaded clothes, yet *something* gave me away and her head swung round and I was pinned to the oak tree by her double-barrelled, black-eyed stare. So I had the answer to my own question – how long before she would see me? – and that answer was about twenty seconds. The fact is that despite all the care I routinely take in such a situation, the deer is better attuned. She is of the woods and I am not. Each time I come back here, or somewhere like it, I have to re-acclimatise, shed influences from beyond the woods that – inevitably – I bring with me. And I am willing to bet that she will have known that I was here long before she trusted my stillness enough to cross the clearing and pause in that last scrap of evening sunlight.

A lesson falls into place here, a kind of first commandment and it cannot be repeated too often. It is that in all our dealings with wildlife, from watching a roe deer cross a clearing on a summer evening to reintroducing lost species (say wolf, beaver, sea eagle) with consequences that will endure for hundreds of years, it always takes time. And mostly giving the required time to watch and learn and understand – letting nature take its course – is where our side of the bargain falls short.

The doe turned from side-on to head-on, and as she did so that low and intense sunlight travelled across her as she realigned head and neck and chest and back, until only her face and chest were brightly lit, lending her a faintly ghostly air. I wondered why she had turned like that. What I expected was that she would turn to face away from me then turn her head back to look at me down the length of her spine, ready for flight. But this had the air of something more confrontational, or perhaps just more contemplative, weighing up my threat, or absence of threat. Then she stamped one forefoot. A soft thud like a muffled drumbeat. Was that for my benefit, an invitation to leave the premises? Or else, a signal...?

A juniper bush shivered and her fawn materialised from its dark-green fleece and came to stand alongside her, a perfect miniature mirror image except that it lacked the sunlit patches, being completely in shadow. The fawn could have been no more than two months old, yet she had obviously acquired from her mother the art of the disconcerting stare. The penetrating effect of those four unblinking eyes pinning me to the tree amounted to rather more than the sum of the parts. The watcher watched – it is a circumstance that always unnerves and I never understand why. Was this not exactly what I sought to achieve? An engagement with nature on an equal footing, creatures of the same landscape.

I studied their double act. What were they thinking? Were they thinking at all? Then, without a sound, they both turned to their right as if it was a well-rehearsed manoeuvre, walked beyond the sunlight to a more open patch of ground, a distance of no more than twenty yards, and without another glance they both started to browse. The doe constantly raised her head to look back but I gave her nothing, no cause for alarm. I was oak-tree still. I hoped that the warmth and the admiration I felt for her and her offspring in that place at that moment would transmit, that she would catch the sense of it and relax.

A memory slipped into my mind. Alaska, 1998, Kodiak Island, thirteen grizzly bears in a quarter-of-a-mile-long arc of river. My guide, a benevolent giant of a man called Scott Shelton, advised me that there is only a problem if the bears pick up stress and anxiety from you. It makes them edgy.

So I had asked: "If you come in a spirit of warmth and admiration, can they pick up that too?"

He smiled. "Yes, of course. Of course they do. That's how I go to work."

These are the kind of people from whom I have learned how I should go to work, people whose paths I have crossed and for whom I have felt profound gratitude.

As I sat watching, and the deer appeared relaxed and the sun faded from the clearing and a bluish tinge of dusk began to infiltrate it, my mind went wandering off along an old roe deer path that had led to a similar encounter, the memory of which was initiated by that miniature mirror image when the fawn came and stood beside its mother. That old encounter had also involved a roe deer and her fawn, and what happened was this:

The two deer were ambling quietly, the doe about ten feet lower on the hillside than the fawn, which walked on a parallel contour. The fawn's ears flicked suddenly forward, her head angled sharply to one side, and in that attitude she froze. For the moment, she could see what her mother could not. But the doe had seen the fawn's reaction. From my position in the edge of the trees and about twenty feet below the doe, I could see the cause for concern now that the fawn's sudden change of attitude had drawn my attention to it, too. A stoat was running easily along the deer path towards the doe, which was hidden by the lie of the land. First, the stoat saw the fawn above him, all eyes and ears. It stopped dead. It stood, in the manner of all mustelids, on its hind feet, brandishing its curiosity.

The doe came on cautiously, nose a-twitch, ears working. The stoat dropped to all fours, unaware of what was just beyond the rock by the next bend in the deer path. It ran on, stopped, stood, eyeballed the fawn again, dropped to all fours yet again, hesitated, stood again. Then in a kind of decisive frenzy, it hurtled round the rock only to meet the staring eyes of the doe a yard away. The doe's front legs splayed wide and her head went down (I defy anyone not to think of Disney at that moment). The stoat effected an emergency stop and reared onto its hind feet once more, all in a single seamless movement, a singular feat of reflex and co-ordination. For two seconds, perhaps three, it took stock of the sorcery at work on that hillside. One moment it had ducked down past what it would presumably recognise as a young deer, the next the path was blocked by its hugely magnified mirror image (and *that* was what had cued in the memory of the encounter). Its response was to hit the rock, which was about eight feet high, and to scale it at a flat-out run. It was an impressive little feat of free mountaineering, but it had the unintended consequence of depositing the stoat by the ankles of the fawn. The stoat performed the nearest thing to a fast double-take I have ever seen in the wild, a two-footed, spine-stretching assessment of first the fawn and then the doe. Then, I think it saw me, and I was its final straw. It plunged into deep undergrowth and vanished from the face of the hillside. Both the deer watched it vanish. Then the doe also saw me, barked softly to the fawn, and they stepped off into the deepest shelter of the trees.

◉ ◉ ◉

Back in the summer wood where this all began, doe and fawn still browsed, inching slowly across the bottom of the clearing. I watched them for a few minutes more, and if I am honest I envied them their place on this wooded hillside, for they had an involvement with it, an intimacy with that land and everything else that lived and breathed there, an intimacy that is light years beyond anything I might ever hope to achieve. Then they moved off silently into the trees. With their disappearance and with the embellishment of the entertaining memory of the deer and the stoat incident, I considered what I had just seen and the dilemma that it had just dropped in my lap. And this is part of the nature writer's condition, for in my head were the words of an old friend called Don MacCaskill. Don was a forester, the Forestry Commission's chief forester for the Strathyre Forest, and a very untypical one at that. He was a brilliant naturalist, an award-winning wildlife photographer, he hand-reared foxes, his dark-room walls were lined with photographs of wolves. The publication of this book coincides with the twentieth anniversary of his death, and had he lived he would have been 100 years old. The words that made me realise, quite early on in our friendship, that he did not quite fit the mould of a Forestry Commission chief forester, were these:

"A forest is not a forest without deer."

And of course he was right. The corollary of that is also true: a deer is not a deer without forest.

So, the dilemma is this. My country is overpopulated with deer, both roe and red, too many for the good of the land, for the good of all nature, for the good of the people who live amongst them, and for the good of the deer. Forest regeneration and replanting, especially of native hardwood trees, is not possible without either drastic culls of deer numbers or thousands of miles of deer fencing that is ugly, intrusive, damages wildlife, infuriates walkers (and nature writers) and is subsidised by the public purse. The only people who like it are the estates that rely on shooting deer, because it keeps the deer away from trees and out on the open hill, where they are much easier to shoot. The phenomenon of what results is called (with vicious irony) a deer forest.

It is wrong-headed in every possible way, not least because both the roe deer and the red are forest animals. It's where they belong. The red deer's banishment to the open hills has created a diminished race by comparison with their European forest-dwelling counterparts, which are a third bigger. All forests should be deer forests. But the first thing that happens whenever a conservation body decides to restore or recreate a native woodland that has been wrecked by overgrazing is that they hammer the deer numbers and throw up a fence. Then the new trees have a free rein and after twenty years, behold the new forest! But a forest is not a forest without deer.

What Don did not say that day was that a forest is also not a forest without the deer's natural predators, not least because we have come to rely these last 200-and-something years on the deer's *unnatural* predator – a man with a gun – to keep their numbers in check. The evidence of your own eyes tells you that a man with a gun is a hopelessly inadequate and hopelessly unskilled means of keeping deer numbers in check. The landowning fraternity has friends in high places. Any number of attempts by the Deer Commission (Scotland), and Scottish Natural Heritage and their predecessors, to compel landowners to reduce deer numbers to the required level have failed utterly. The required level means a deer population that is low enough to permit the land to flourish, to permit native woodland to prosper, to permit every native habitat to restore and expand, to permit every scrap of biodiversity that the land is capable of sustaining.

In January 2020, one more attempt was made to pressurise the Scottish Government into drastic action. It came in the form of a report by Scottish Environment Link, a coalition of eighteen organisations representing environmental, community, wildlife, forestry and outdoor recreational interests. It proposed "...a few simple steps...such as statutory regulation to ensure deer densities are reduced to sustainable levels in every area; phasing out public financial support for deer fencing; broadening participation in deer stalking to involve local communities..." And all of this

"...in the light of the accelerating climate and biodiversity crisis..."

I fine-tooth-combed the report for a hint of a vision that went beyond the man-with-a-gun syndrome. I found none. I looked for some acknowledgement that wildlife is better at managing wildlife than we are. I found none. I looked for a single mention of the red deer's natural predator – the wolf. I found none.

The thing about the wolf is that it doesn't just kill deer, it keeps them on the move by its very presence, and that is fundamental to deterring the tendency of huge deer herds to browse a single area of vegetation to the bone, creating landscapes where trees haven't got a hope in hell. Instead, the deer must move on at the rhythm demanded of them by the wolves. Guns are not much of an option in a forest, but most wolf packs are born to it. Conservation thinking in Scotland is still fearful of wolves, or at least it is fearful of public opinion because public opinion also equates to the membership of conservation organisations. Yet public opinion will change once the wolf is here (if we ever pluck up the courage) and it learns what a wolf is actually like to live alongside, how it really goes about its business. The way to learn about wolves – about any animals at all – is to live with them and watch what they do. Reintroduce the wolf and everything in nature starts to make sense and biodiversity follows in the spoor of the one creature that actually understands how to make all the rules, because that is what it has always done. It is as simple as that. It really is.

I have strayed a long way from the moment when I watched the doe and fawn slip out of sight, and I remembered Don MacCaskill's words. My dilemma is not that different from that of anyone else who likes watching wildlife. It is that watching deer is as intriguing as watching otters and watching foxes and watching pine martens. And a roe deer mother and fawn in a summer woodland clearing is a sight of quite undeniable wild grace. Shooting something like this from a high seat or a hide is frankly grotesque and teaches the animal nothing. If the gunman happens to be an expert shot, it is argued that the method of killing is "humane", as if that was a good thing, a curious conceit for the species that is the most destructive killer of wildlife on the planet. If the gunman is not an expert shot, then the chances are his quarry does not die instantly but slowly and in agony.

The relationship between a natural predator and its natural prey is that the prey species understands the predator, understands what it is and what it does, evolves strategies to try and outwit the predator. This means, in turn, that not only is the prey species diminished in numbers, but the many survivors are healthier because of the very presence of the predator. That, too, is a simple equation.

◉ ◉ ◉

I watched the last of the light fade from the clearing. I have always loved that hour when daylight turns to dark, especially high-summer dark, which, on the Highland Edge means a bluish grey half-light. Eventually, just as I was about to leave, the unmistakable voice of the roebuck boomed into the clearing and hung on the air, deep and throaty, two harsh syllables designed to put the fear of death into any other roebuck within earshot. Fear of death is only a slight exaggeration. Few creatures the length and breadth of wild Scotland defend their territory as uncompromisingly as a mature roebuck in high summer, for that is also the roe mating season. I watched his silhouette follow precisely the line taken by the doe a few minutes before. She was somewhere out there just ahead of him in the gloom, but he knew precisely where. There is a dead oak tree, a runt, not far away, and stamped into the grass around that tree is a perfect circle, the ring of the roe mating season where buck follows doe in a ritualised merry-go-round of salacious intent. I had seen him earlier in the day as he crossed a patch of brilliantly sunlit bracken and he was a beast in his prime and lord of his manor. There are places in the Lowlands where roe deer numbers are a problem, but on the edge of the mountains and in woodlands like this you do well to get a view like that. But the glimpse of his silhouette as the gloaming gathered around the edge of the clearing (for he had paused where she paused, scenting her progress), *that* was the prize, that was the glorious moment of confirmation of Don's words, and what a photograph he would have made of the moment. I simply kept it in my head until it was time to write it down.

A forest is not a forest without deer.

And the deer is not a deer without the forest.

Never was. Never will be.

Chapter Five

Inside the Arc

SO OFTEN IN RECENT YEARS, summer has begun for me at Flanders Moss. The fourth of June 2019, and the arc of mountains from west-north-west to north-east finally turned summer blue. So much of my day-to-day inclination to keep nature's company – from planned day-long ventures with a destination and an objective in mind to spontaneous grab-a-notebook trips to who-knows-where and let's see what unfolds when I get there (the latter option always produces the best days, always!) – so much of all that unfolds between the valley of the upper Forth – the Carse – and that arc of mountains. Inside the arc, then, is where I go to work most often, this country in the heart of Scotland where Lowland meets Highland. I think of it as the Highland Edge, and this is how it rubbed off on me through the months of what we used to think of as summer.

The profile of those mountains reassures with its familiarity, its proximity, its neighbourliness. From the tree-top-level deck of the timber observation tower in a corner of the Moss's national nature reserve, and on such a day of summer blue, I am apt to recite the mountains names sunwise across the bottom of the towering sky, a roll call of the ever-present: Ben Lomond, Ben Ime, Ben Narnain, Ben Venue, Stuc Odhar, Ben Ledi, Ben Vane, Beinn Each, Stuc a' Chroin, Ben Vorlich, Uam Mhor. They had begun the year that pallid shade of midwinter-mountain-grey that is only occasionally enlivened by the white of snowfall, but nothing like as often as when I first got to know them forty-something years ago. Winter has lost its reputation for reliability in the first and last mountains of the Highlands, lost its edge, its bite. And having given space on the page for that thought, the same could be said for all four seasons. If there is one overarching theme that characterises these last five years I have been consumed with an intense scrutiny of the march of twenty seasons, it is that they no longer deal in seasonal norms. The Met Office's TV forecasters announce on the day before the first of September, December, March and June that "tomorrow is the first day of meteorological autumn/winter/spring/summer". It is as well to advertise the dates, I suppose, if only to remind us of a time when the seasons more or less behaved the way the Met Office would presumably still like them to behave, but its job has not been made any easier by the havoc we have wreaked on the climate.

On repeated visits here I watched the mountains sluggishly morph out of winter from slushy grey to sloppy green, and given that, to my mind at least, green never suited a Scottish mountain, the particularly unenthusiastic shade of green that spring mustered suited them even less. A few sloppy green weeks drifted by, lethargic as a newly woken hedgehog, but then out of nothing at all, that day of June stillness dawned, and with the sun not yet risen the mountains smoked an agreeable shade of purplish grey. The rising sun made a curiously garish pink of that for a spell so brief that you might wonder if your eyes were tricked by nature. It lasted only as long as the sun took to crack open the eastern horizon, shake itself free from what felt like an eternity of hibernation but, in reality, was only its brief night underground. Then it became airborne and then the blue began, the palest insinuation. It would spend the morning consolidating, so slowly that it can seem unchanging, but as the sun climbed and brightened and travelled round that rarest of weather phenomena in Scotland that is a cloudless sky, and as the hours drifted by in a daze of contentment, the sorcery proceeded imperceptibly. By early afternoon I would finally be convinced that it was summer, convinced by the depths of blue in the mountains. It was still a pale blue by the standards of nature's palette that accommodates everything from forget-me-nots to the navy of a starry sky that declines to blacken. But it was my true shade of summer mountain blue the whole length of my neighbourly skyline. The new season was at hand.

Meanwhile, and in startling contrast both to its workaday self and to the blue mountains, the Moss was clothed in a surreal shade of bridal white. The effect, even in June sunlight, was eerie. Such was the impact of the dense growth of bog cotton – cottongrass – that a breeze making mischief there contrived the illusion of a swaying, dancing sea. Whenever the breeze dropped, the feathery heads of every one of millions of individual stems took perhaps half a minute to grow still, by which time the breeze had resumed from a slightly different airt, and the dancers and the swayers jigged and reeled into each other then settled again into something like a slow, coherent rhythm that became briefly oceanic.

The tremulous nature of every square foot of the entire foreground-to-middleground of that Lowland-to-Highland landscape was further enlivened by hundreds of knee-high birch saplings that waded gleefully out into the white sea and its peaty shallows. Nature keeps trying to reinvent the Moss as a birchwood, and doubtless has been trying for centuries, ever since this stopped being the bed of a sea loch (archaeology has unearthed sea shells here, whale bones). But because it is classified now as a raised bog and managed by Scottish Natural Heritage as such (for there are few rarer habitats in Scotland and this is the largest example, by far), that means keeping the water table high. So every now and then the birch saplings are cut down. And I'm not too sure whose side I'm on with that particular land-management dilemma.

Down from the timber tower to wander the partial boardwalk footpath and the grown-up birchwood that flanks the Moss to south and east. Quite apart from the suddenly ubiquitous presence of lizard, dragonfly and damselfly (see the Prologue if you didn't begin at the beginning), you are now down at the sodden level of pondlife. In a summer as wet as this one would turn out to be, the feeling is of a vast and unsubstantial tract of the raw stuff of the planet, a thing more of water and moss than anything that you might call dry land. But that early June day was all sunlight and warmth, and given that the water and the water plants live here in abundance, the lizards (splay-footed, twenty-toed, pyjama-striped along their entire length, elegantly curved, the stripes serving to accentuate the curves) and the dragonflies/damselflies (I have invented my own shorthand – dragselflies) cross your path every few yards. Both tribes ask nothing more of life than the bare-bones-simplicity of the Moss's hospitality.

Then there are the birds. Some you will see but hardly ever hear (redpoll, hen harrier), and some you will hear but hardly ever see (cuckoo, cuckoo, cuckoo, cuckoo). Every bird call on that day of days was heightened by a curious effect I had never noticed before: it was as though they had been lit up and brightened by an acoustic layer of non-stop larksong. I had not noticed it on the tower, but now at ground level, it lay like a dome over everything. The nearest bird when the phenomenon stopped me in my tracks was not twenty yards away, not fifty feet up, but by focussing on that single lark I could hear the others higher and further off, rising, pausing at the zenith, on top of each impossible, slender column of song. Listen hard. How many? Twenty? Thirty? Fifty? And how many more out of earshot across the soft white gentle throb and sway of the whole enchanted square mile? It even made two cuckoos sound beautiful. Ravens were gargling in the backs of their throats and made it sound like music. The thin drip of noise that is the reed bunting's signature was as vivid as a melting snowflake snared in a juniper bush. Stonechat (which is so melodious the nature reserve staff on the island of Rum used to refer to geologists as stonechats because of the noise of their hammers on stone) and sedge warblers (I fondly imagine some old academic sage unearthing a Gaelic manuscript in the bowels of St Andrews University which reveals a sixteenth-century name for the sedge warbler that translates as "the insomniac haverer of the reed bed")...even these were made glorious wrapped in larksong made symphonic.

◎ ◎ ◎

Five years ago, walking up this hillside path in Woodland Trust Scotland's life-affirming Glen Finglas estate, I came face to face with a green woodpecker in circumstances so remarkable that they have come to define for me the entire enterprise of a quartet of books, as they seem to have done for many of those who read the first of the books, *The Nature of Autumn* (Saraband, 2016). The encounter occurred on the first day of that old autumn, and it remains among the enterprise's most enduring images. It came back into my mind in all its glorious technicolour as I neared the tree from which it staked its claim to fame, and the recollection disturbed the relatively prosaic thought process that was preoccupying me at that moment. It was a bit like one of the Hums of Pooh that, you may remember if you are a Pooh fan (and who isn't?), "came suddenly into his head, which seemed to him a Good Hum, such as is Hummed Hopefully to Others". Mine wasn't as good as one of Pooh's Hums, being only a single six-syllable line when it was interrupted, but I thought of it as a kind of mantra. Once I had ordered its four words into their most effective rhythmic flourish, I decided it might be the beginning of a poem, if not one that might be Hummed Hopefully to Others:

"Birch, willow, alder, oak," was how it went, each word timed to coincide with a footfall, so that I marched to the beat of my own drum, or at least walked to it, for I tend not to march in nature's company. What had brought it back into being was the transformation that trees (and these four species, in particular) had wrought on the path even in those few intervening years; trees that leant out into the path's airspace from the forest edge, narrowing its shoulder room, greening my summer passage, every tree stretching, broadening, burgeoning with wellbeing.

It came to me then as I approached a certain partially obscured oak (even more obscured now than it was then) that it was the woodpecker tree; five years ago I had written that it had

> *...just moved in an eye-catching, instinct-rousing way. Eye and instinct fused to form my one indispensable life-support system, my all-season-all-weather shadow. I never go out without it and it is forever muttering in my ear...*

What followed was one of those moments that simply go beyond, that touch rare depths of communication between nature and nature writer, and if you subscribe to the ideology articulated by the American architect Frank Lloyd Wright in the words "I do believe in God but I spell it Nature" (and I say a heartfelt "Amen" to that), then God-given is as fine a way to describe it as any. I am more inclined to think of such moments when they do occur as nature tapping me on the shoulder so that I pay particular attention. "Watch this," is what nature is saying.

I stopped briefly, trying to remember exactly where I had stopped five years before, but such was the transformative growth of this young forest on both sides of the path that I recognised almost nothing, except the oak that had spilled the green woodpecker and left me spellbound and enthralled. You had to be there. Or you could just read *The Nature of Autumn*.

"Birch, willow, alder, oak. Birch, willow, alder, oak."

The mantra resumed, and I walked on to its agreeable rhythm. No woodpecker flew, not today, and I was relieved in a way, for that other day five years before is so far out there on its own that no reprise could ever bear comparable fruit. Instead, I contented myself with the knowledge that such moments are possible when your species and mine allies vision to sympathetic thought and action on the ground to give back to nature that which was ungraciously removed by our predecessors. As a species, we have not yet completely lost the sense of our place in nature's scheme of things, nor the sense of our obligation to her cause, although on a global scale that fealty is frail and fading.

We claim to know the worth of what we call conservation now, so we can never again claim that we didn't appreciate the toll of our wrongdoing. Crimes against nature are crimes against everything, because in the final analysis, nature is all there is. From the historic obliteration of our wolves to today's grouse moor's toll of eagles, to the wilful destruction of the Amazon rainforest, to the catastrophic melting of polar ice and the "third pole" of Himalayan glaciers, to the incomprehensible greed and indifference that stalks elephant, rhino, lion, tiger, whale: all these do a gross disservice to the whole planet and everything in it. When John Muir wrote that "when we try to pick out anything by itself, we find it hitched to everything else in the universe..." he was far from the first human being to entertain the thought, merely the most celebrated – that and the fact that his writing is just so damned quotable (a mixed blessing, for we are often ignorant of the context of his best one-liners). It was an idea well understood for centuries by populations who lived hand-in-glove with nature – from the circumpolar tribes of the Arctic rim to the African Maasai and Australian aborigines, among many others – long before the influence of European colonists corrupted all their lives. If you think it takes a particularly athletic imagination to leap from John Muir to the Dreamtime to Glen Finglas in the summer of 2019, consider the following:

One – nature was poorly served in Glen Finglas before Woodland Trust Scotland acquired it in 1996. The removal of spruce plantation and the pressures of overgrazing by sheep and deer was followed by the planting of a million broadleaf trees.

Two – nature galvanised at the removal of the twin strictures. So now two wholly negative processes were replaced by two positive ones. This is what I like to think of as the forgiveness of nature. The treeline creeps upwards. Wildflowers thicken gloriously and flourish knee-high on Stuc Odhar, Ben Ledi's mountain-shaped little sister to the west.

Three – small birds and mammals, bugs and butterflies flock to the change. Warbler song is as incessant as wind; the trees rock with it. Voles flicker on the bare places along the path, a kestrel hangs, a hen harrier cruises a yard above the heather. Bullfinches hang upside down on an alder at eye level; they contrive to be both small and plump at the same time, the deepest pink in the wildwood, the most mellifluous single woodwind note you will ever hear.

Four – larger birds and larger mammals follow them in. Black grouse numbers increase, a rarity anywhere in Scotland. Golden eagles start to hunt over the higher slopes, working right at the very southmost edge of the Highlands. Sea eagles follow them sometimes. What's next? Goshawk?

Five – suddenly, there is something like a trend towards nature in a woodland environment, from Orkney to the Borders. Here, in the Trossachs, Woodland Trust Scotland, the Forestry Commission Scotland and RSPB Scotland have joined forces to create the Great Trossachs Forest, 160 square kilometres of woodland, loch, river, wetland and open country, a continuous landscape mosaic with nature's needs at its heart, and extending right through the Trossachs from Callander in the east to Inversnaid on Loch Lomond in the west.

For several years now, and from a favourite perch with a view to a certain piece of land and water and woodland, I have looked out longingly thinking that here should be beavers. It is inside the national park, and national parks of all places should be at the forefront of species reintroductions, but in Scotland they never have been. I think they are obsessed by tourism, while serious conservation is left to others. Happily, the fruits of those others are all around. It is "others" who have taken the idea of beaver reintroduction and run with it. The official trial in Argyll was a qualified success, and the unofficial reintroduction into the Tay river system (by people who chanced their arm for nature because people who get paid to do it were so lethargic) was an unqualified success, whatever a few loudmouthed farmers with shotguns might say, and it is the Tayside beavers that prosper spectacularly and march westwards towards the Argyll sept of their clan. And now they have begun to send exploratory missionaries south into the Forth river system.

And I heard a whisper about what I had been longingly thinking, and what I heard was that it had come to pass: the beavers have found their own way here. So I followed a certain woodland track to where a river swirls between lochs, until the wetness of the land and the fankling trunks and broken boughs of fallen trees brought me to a standstill. Everywhere ahead and to left and right was either water on the move or heaving uneasily in sluggish pools, for this had been the wettest of wet summers in this neck of the woods and the water table was high. And all that water was hampered or stymied or redirected by the trees it had pulled from the soft earth and which now lay at crazy diagonals or slumped horizontally in the shallows where they drowned. Most of my sightlines extended no more than fifty yards, some much less than that.

I sat on a felled alder that had come to rest at waist height, an agreeable perch for me if not for the tree. It was late afternoon and it had grown warm progressively through the day. Over my shoulder, sunlight blazed on wooded hillside, but this deep in the tree-and-water-world something like a bright gloaming prevailed, a deep-green and smoky gold light hung above peat-black waters and swirled through every shape and shade of tree bark you ever saw. I stilled, stared, listened.

Water sounds: the mainstream of the river, out of sight but no more than fifty yards away, muttered with that soft deep bass undertone rivers give voice to when the channel is over-full; nearer at hand, the higher pitched slap of the eddies of the overflow among tree trunks both standing and fallen and neither one nor the other but propped up in improbable diagonals that create vee-shaped reflections where they emerge from the water; and a muted gurgle where some underwater impediment generated a swirl, a gentle whirpool.

Then two things at once, a patch of white on the trunk of one of those propped up diagonals, a lichen-encrusted birch. Just as the binoculars focussed in on it to reveal the vertically cut axemanship of beaver teeth, there was the simultaneous splash of something somewhere deep among the waterlogged trees and between where I sat and the river. At once, there was a scatter of half a dozen mallard ducks I simply had not seen, presumably because they had been at rest on a bank out of sight when their tranquillity had splintered apart. There were two possible explanations. One is that they had simply sparked off a fight among themselves, which is by no means out of the question. The other is that a beaver, or beavers, had appeared too close for their comfort, they had roused at that appearance, and a beaver tail had thrashed the water – the characteristic alarm/discouragement technique that is the typical first response of a territorial beaver with a point to prove to something of which it disapproves. The ducks made an ungainly procession up through the tangle of branches until they had clear air to work with, then they headed for the river.

Silence, effectively deepened by the water sounds.

I felt a new tension, a phenomenon I suspect had been created by my mind and nothing else.

The trees gathered closer. Ditto.

And yet I did not imagine the deepening connection with the here and now, the landscape, nature letting me in a little more. Another look at the white patch on the angled birch trunk, suddenly illuminated by a thin shaft of sunlight: the bark had been removed from the topside and the flank facing me. The grooves of beaver teeth are very clear, vertical and parallel and so close together they often overlap. I tried to visualise how it was done. I know that beavers can and do climb trees, which may seem bizarre given how efficiently they can fell them. But the beaver that gnawed bark on that birch must have been standing on the canted over trunk directly *above* the patch it chose to eat; and to do that, it had somehow managed to generate an airy launch vertically upwards from the water over which the trunk now hangs. I have trouble envisaging the manoeuvre, or fathoming why it had to be that patch of that trunk.

And I suspect that it had been disturbed in the process, because it had gnawed an area no more than a couple of square feet, which is a small reward for the nature of the endeavour. Or did it just fall off into the water from what must have been a precarious stance. Looking at the teeth-marks, which were all vertical, the beaver either had to be standing on top of the trunk and bending over to reach the bark, or it had to be clinging to the top of the trunk, suspended by the claws in its front feet, neither of which is a comfortable proposition. Nothing explains it, except that beaver behaviour is erratic, inconsistent and unpredictable, and our notions of logic have nothing to do with it. But it was an intriguing introduction to the presence of beavers in the Trossachs.

I considered all this in the lull that followed the unexplained splash and the mallards' flight. But then there was another splash, then another, then another…then the sounds of watery mayhem. The still water at my feet began to rock gently. As I waited and watched I found more evidence of beaver activity: a felled tree, a branch in the water gnawed to a perfect pencil-point shape, bare patches where bark had been removed, and all of them bright white and fresh, and all the while the splashes resounded.

Another hour. Nothing revealed. Silence effectively deepened by water sounds. I stayed until the tension eased. Until the frustration ebbed, to be replaced by a kind of benevolent glow, the revealed certainty of beavers in this land where they so obviously belong, and where I had longingly thought them into being.

◉ ◉ ◉

So that's what it means when John Muir scratches his bearded chin a hundred and more years ago and then takes up his pen again and writes, "When we try to pick out anything by itself, we find it hitched to everything else in the universe." You look at a landscape, you see its pockets of old native trees clinging on in the face of misguided land use and ignorance, then the vision comes and the landscape is reborn, effectively in its own image, the image of how it grew and evolved when it was left to nature. And with the new trees, the new flowers, the new and untold possibility, others come, and a beaver has wandered from Loch Earn through Glen Ample into Loch Lubnaig and into the limitless potential of the Forth river system, and because beavers know how to find other beavers (they follow the spoor, they read the signs, they scent the air and they know that everything is hitched to everything else), they eventually meet in a revitalising wonderland of native trees and water on the move the way they like it, and they celebrate the meeting in its most secluded grove of living and half-living trees and water they can manipulate and enliven. They, too, are hitched to everything else in the universe, just like the old trees and the new trees and everything else that followed, that crossed the hill or swam upstream to the summons of nature. And all this happens, is happening right now, inside the arc between river and mountain, and I come again and again to drink it all in and write it down, because in my own way, I am hitched to it, too.

I had wandered away uphill from the woodpecker tree and the mantra had faded from my mind (I never move at the same pace for very long). Instead, I tuned in to ravens, which maintain a constant four-season presence at Glen Finglas. By midsummer, the young are articulate flyers (they are the earliest of nesters) and halfway to articulate in raven language, which is much more difficult, as rich as Latin and much, much older. There were two ravens, unseen for the moment, screened by the press of young trees elbowing each other for shoulder room. They were calling to each other in a way I hadn't heard before, but to my ears ravens almost always have something new to say, given their vocabulary, and – it seems to me – their regional accents. These two each repeated a single note and more or less constantly, the voices a perfect third apart, but never once did they coincide their notes, so never once did I hear a note of perfect raven harmony.

I wonder about this, about how a lot of those wild creatures that produce notes we might think of as musical seem to avoid harmony. Whether that avoidance is deliberate or not, I don't suppose there is any way of finding out. The most obvious manifestation is in wolves: they howl in discord, and if two animals accidentally slip into harmony, one of them slips out of it at once. The mellow voice of a whooper swan becomes cacophonous when a new skein flies in to land among settled swans on the ground. And here were two ravens crying their own single note but staggering the intervals of their calls so that not once in ten unbroken minutes did they overlap. Late in the day, and high enough up the hillside to look out over the born-again woodlands of Glen Finglas and far into the reserve's woodland neighbours, I watched a group of eight ravens together, all of them calling as they flew and back-flipped and side-slipped. Now the calls overlapped all the time but although they ranged over two distinct octaves, I could detect nothing at all that sounded like harmony.

The greatest challenge in the field of natural harmony is surely the humpback whale. In the late summer of 1998, I stood with around a dozen other humans on the deck of a small boat surrounded by humpback whales. The skipper announced he was cutting the engine and putting a hydrophone over the side, and uncommanded, every one of us fell silent. Whale song has a range of seven octaves. I don't know how many were covered by the sounds that drifted on board through the tannoy, but it was most of them. Four, five, possibly six whale voices, Ellingtonian in their reach and structure and developed variations on recognisable themes. Baritone sax to piccolo and beyond.

Wolf, whale, swan, for me the three most thrilling voices of land and ocean and air, but none makes any attempt to dovetail an individual voice to any others. And it was while I was in Alaska that I learned of a tradition within some of the tribes of the far north of both Alaska and Canada that it was Raven that made the world. And one of Raven's many gestures involved a particularly sad wolf on a beautiful moonlit night.

Raven asked the wolf why it was sad, and the wolf said he wanted to talk to the moon but didn't know how. Raven asked the wolf why he didn't howl to the moon and the wolf said he didn't know how to howl. So Raven changed himself into a wolf and showed him how to howl.

I never heard a raven howl, but I have heard that octet range over two octaves, and I have heard that pair hit those same notes again and again and again and right in the middle every time, and never once tried to make them harmonise. And the ravens in my sky in the summer of 2019 reinforced what I also learned in Alaska twenty-one years before (and from a Canadian university biologist), that ravens have a vocabulary outmatched by no other northern hemisphere creature apart from us. Listen to them sometime, really listen, eavesdrop on the conversation of a sky-scribbling squad of ravens on a summer's day and be amazed. As you listen, you will have another essential truth reinforced for you: that when you try to pick out any one thing by itself, you find it hitched to everything else in the universe. Which is why I didn't bat an eyelid the day I learned that Raven taught wolves how to howl.

⊙ ⊙ ⊙

Late in the day, driving slowly through the flat fields of the Carse of Stirling and watching a sunset-smoked sky settle down into purple and grey above an arc of mountains that had paled to no colour at all that I could name, a brown hare broke cover from beneath a roadside hedge in front of my car. But instead of seeking refuge back in the field it had just left or through an open gate in the hedge on the other side of the road, it opted to lope along the road about twenty yards in front of me. The road was single track and deserted apart from my car and the hare, and I slowed right down. We must have covered a hundred yards that way, at about fifteen miles per hour, which is a canter for a hare. Finally, it stopped in the roadway where the hedge ended and an open wire fence began. It slipped gently under the bottom strand, and resumed its unforced lope through thick grass. I stopped, wound down the passenger window, reached for binoculars, focussed just as the hare stopped and sat back. As it sat it raised its ears as tall as hare ears will go, and in that light they were jet black. They looked like nothing so much as a pair of raven feathers. Like I said…hitched to everything else in the universe.

Part Two

Song for an Unsung Shore

Chapter Six

Solstice

THREE IN THE MORNING. It doesn't take much to waken me at such an hour. I am not the most efficient of sleepers. An unaccustomed bed in an unaccustomed house is as likely a circumstance as any. The unaccustomed window with its unaccustomed view is open. There is light through a gap in the curtains. It is the week of the summer solstice. Cross the room to the window: the sea is across the street (the high tide will pause at slack water a few yards from the pavement). Its colour is a phenomenal shade, the palest turquoise I ever saw, a shade surely unique to a pre-sunrise east coast sea. But the sky in the north-east looks ready to burst into flames. By 4.30, the sea will dazzle. In the course of such a day it will scatter shades from that turquoise to mid-grey to pink to colour-free dazzle to slate blue to royal blue to a late afternoon tapestry of blues that will drape navy along the horizon, then to slate and lilac and through paling greys to white at about 10p.m. And in twenty-four hours' time it will work its way up to dazzle again and repeat for seven unbroken days.

It was to be that kind of week.

◉ ◉ ◉

I find that the years pull me east more often. Something about the pared-to-the-bone simplicity of that coast and the lure of a quiet country: land a curve of dark red sandstone cliff, sea a slab of smooth and blue-grey unpolished marble untrammelled by islands and undaunted by mountains. The sun will rise out of such a sea and so will the moon. Have you never seen the sun rise out of the sea? The moon? Oh, it really is quite fine. Uniquely among all the landscapes of Scotland, summer is the east coast's finest hour. And compared to the west and the north, tourism is sparse; its shore is unsung. I speak, I sing, as a native. The east is my native shore.

On the west coast, that sliver of my inheritance that is Celt lays claim to me; I succumb to a thraldom of Hebridean islands and their complicit mountains. I stir to the summons of the Skye Cuillin, the subtler seductions of Cìr Mhòr on Arran, Ben More on Mull, lowly Dun I on Iona, Beinn na Gudairean on Colonsay, the Sgurr on Eigg, Dun Caan on Raasay, Clisham on Harris. On the north coast, Orkney's pod of whale-shaped islands speaks to me in calmer tongues, more George Mackay Brown than Sorley MacLean, whereas Shetland and Fair Isle are where (it seems to me) the north of the world begins and I thrill to that benevolent shiver of Norse-tongued names for island, hill, rock and bird (dunter, not eider; shaldur, not oystercatcher, swaabie, not black-backed gull; tirrick, not tern). An exquisite northern other-worldliness begins there and sometimes I have aspired to that...Alaska, Iceland, Norway, beyond where the Arctic increasingly lurks just within reach of my idea of what may yet be possible.

All of that began aged about twenty. But when I look on the east coast, something else happens, something that laid claim to those first twenty years before islands and West and North knew anything about me, and all I knew of them was in books by other people, glimpses on family holidays. And, of course, the first twenty years of anyone's life lay down the bedrock on which whatever follows must stand or fall, for good or ill. And I was lucky with my bedrock. Dundee, on the north shore of the firth of the finest river in the land, sculpted from two hills of its own, was more surely bedded in the county of Angus then; "Dundee, Angus" was its postal address. And forby the hills and mountains that crowd the north of that county, forby the beaches and the bird hordes of the firth, there was the tantalising coast of my young years that reached from Montrose in the north by way of Auchmithie, the Bell Rock, Tentsmuir and Fife Ness to the East Neuk and the Isle of May. That was the seagoing inheritance into which I was born, and its light is all the shades of gold.

And if you are sufficiently well attuned, oh how it sings! It is Scotland's skylark coast, for the fields and moors roll down to the clifftops along much of its length, and the larks spring from the grasses and clusters of wild poppies and ox-eye daisies, and by the time they are a yard off the ground they are singing, full-throated from the first note, as self-confident as Beethoven's *Fifth* or Louis Armstrong's *West End Blues*. And in places they cling to their slender columns of song held aloft on warm summer winds a dozen to an acre. And the east coast throbs with the croon of eider drakes, and wherever those grasses and poppies and scrubby bushes and small trees spill over the edge to thicken the sea cliffs, summer is further adorned by carolling whitethroats and the grin-inducing och-aye of puffins.

Such is the land into which I was born, where summer announced itself in shades of red sandstone and tawny sand, seas the width of the known world and skies that went on forever. I astounded my parents at the age of two or three when I said "kittiwake" without any prompting: I heard it, I said it. Whenever I revisit the east coast now, I feel something akin to what the writer and mountaineer David Craig (sitting contentedly on a rock ledge) called "the sense of rightness regained", and there is nothing the finest views in all the sung shores of the world can do about that. It means, in my case, regaining the native rightness, the landscape whence all the others sprung, the original understanding of my place in nature's scheme of things (tiny, lowly, watchful, wonder-filled). There is also a revised sense of belonging, which seems to matter to me more at least some of the time, and more often than it used to.

And there is, of course, much more to the east coast than the portion that accommodated my boyhood and youth. Much of it is essentially Lowland in character, but a Caithness coast, for example, is a very different place with a very different sensibility. More often than not, until fairly recently, questing travels on that coast were inclined towards north, and it was Neil Gunn's eyes I looked through. But then this book's journey took me to Berwickshire and the cliff-hunkered village of Burnmouth, and there, over the course of a single week, I learned to imbue the east coast with a music that I had only heard dimly until then. For the

Berwickshire coast itself was also something of an unsung shore in my own mind, a portion of my native shore that mostly lay beyond my accustomed horizons. It seemed to me then that as I was writing the final volume of this quartet of the seasons, I would benefit from immersing myself in a comparatively unfamiliar tract of my native shore to sit alongside the more familiar. Then I stumbled across a cottage a handful of miles from the Border and with the sea on the other side of the street. And the first thing I heard when I parked and opened the car door was the ever-so-slightly eerie, atonally beautiful, curiously wolf-like chorale of grey seals.

At that moment, even before I had unpacked the car, I thought that perhaps I had found a defining song for my unsung shore. A simple principle fell into place that first afternoon and evening, and it grew on me far into the night. It was that the ebb and flow of the tide coincided with the wax and wane of the sealsong. The ebbing tide uncovered acres and acres of flat rock carved into long, parallel north–south furrows, and as soon as the furthest-out rocks began to appear above the waves, so did the heads of the seals. It is pushing it to suggest that they formed an orderly queue to step ashore, but the effect was something similar, for they gathered on the seaward side and swam around until they found what they were looking for. And once they settled (a relative term for a grey seal colony on haul-out rocks, for theirs is a restless way of life), and from time to time and at any time of the day and night, and in concert and solo...they sang.

All week they sang and they sang, but their song was intermittent and so unpredictable that it surprised me every time it resumed. And after a couple of days and nights it did begin to feel like a resumption, as if it had just retreated beyond earshot and I never heard it go or realised it had gone until it drifted back hours later; and in much the same way I would stop hearing the sea for a while (the weather was a midsummer idyll and the sea responded in character so that it lay much of the time like that slab of sunshine-polished blue-grey marble I had grown up with decades before, smoky pale grey unpolished marble after sunset), and at low tide when it had retreated beyond the rocks it was little more than a conspiratorial murmur.

The great joy of that potently harmonised sequence of days and nights was to awaken in the wee small hours with a hint of dawn in the eastern sky beyond the open window and perhaps it was a single seal voice that crept in under the old stone lintel, and I would imbue a single voice with an added edge of wildness for no other reason than the serene mood of the time and place and atmosphere and moment. Then an answering voice would be joined all at once by half a dozen others, and once again it had struck me that they seemed to avoid harmony deliberately, and again I thought of wolves, for that is the way with wolfsong, too.

The view of the land across the bay in front of the cottage ended in a little headland, which, in a certain light, had the appearance of the head of Hollywood's idea of an Apache brave in profile. After a few days it became apparent that the rocks beyond the cliff he embellished were the source of much of the sealsong. The last full day of my stay turned out to be the summer solstice, and the tide table suggested that a couple of hours of late morning and early afternoon sitting on the clifftop directly above the Apache would offer prime seal viewing.

It did. As the sea fell back and back, the water was studded with the heads of approaching seals, and it seemed to my inexpert eyes that they waited for the sea to clear from a particular long and almost level rock before they emerged. Eventually, there were about sixty grey seals, arranged along four rocks, and as it happened the rocks were aligned largest to smallest from right-to-left, south to north.

The inland side of the big rock's platform accommodated two small pools, each about the size of a single adult seal, and a kind of natural waterslide over seaweed. In the course of two hours, these were in almost constant use. I hesitate to say the seals bathed in the pools given that they spend so much time in the water anyway, but it looked that way, or perhaps they liked the feel of the seaweed; and there was a suspicion of glee at the way they used the slide to move from rock back into water, but that too could have been human imagination rather than seal biology.

The singing was more sporadic than ever in the middle of the day, and most of it came from the smallest rocks. I wondered about its purpose. It's not a contact call, for most of the time they are literally within touching distance of their kin. But it is a far-carrying sound (wolves again, and whales!) so perhaps it is designed to send information up or down the coast to other haul-out rocks. Or out to sea. Memo to self: must find out more about grey seals.

Those clifftop hours were as agreeable as any of the entire wild year. But then, just when I thought that it couldn't get much better, there were dolphins.

They breezed across the wide entrance to the bay in speeding, gleaming curves with dorsal fins curved like breaking waves. Sporadically they galvanised into steeplechasing leaps. The sun bore down on them. The seals watched them, or ignored them (they are no slouches at the art of porpoising themselves when the mood is on them). I suddenly realised I was gasping out loud and smiling. Seven or eight dolphins, perhaps more, putting on a show for the landlubbers on seal rock and clifftop.

Then the perfect dive. The last of them soared from the water, steepened its descent, and as it re-entered the sea it was perfectly vertical. In my mind's eye its spread tail flukes hung for a moment and suspended the weight of the entire dolphin directly below it, then followed the most graceful torso of that inshore water back into the sea. That was the image that endured, that became the shape of the midsummer solstice, just as the seals became its song.

But the moment would produce the most extraordinary encore. Later that evening, I was sitting in the cottage living room listening to music and staring out of the window at that sea across the street. By then I was beguiled by it, for it was both constant and constantly transforming, changeless and constantly changing its mind, its voice, the very process by which it came ashore. If your imagination runs to the notion of a slow-motion kaleidoscope, this was it. Wave patterns became virtuosic, colour became relentless diffusion. No matter that the sun was setting 200 miles over my shoulder in the Atlantic Ocean somewhere west of the Mull of Galloway, within the symphonic scope of the North Sea were tiny tone poems of every colour I could think of and some I could never have imagined. A wave a few inches high curved lusciously around folds of black, mahogany brown and egg-yolk yellow before it curved and splintered, collapsed and vanished as if it had never been. In its place another arose, ice-blue, purple, crimson, gone. These lights flickered and faded in the shallows and among the few still bared rocks.

Diffusion.

I looked at the word on the page where I had just written it down. Then on a whim I reached for a dictionary and found that in the context I had ascribed to it, it suddenly amounted to something rather more than the sum of the parts.

> *1. to spread (something) out freely in all directions; 2. to break up and distribute (incident light) by reflection; 3. the process whereby particles of liquids, gases or solids intermingle as the result of their spontaneous movement; 4. reflection of light by a rough reflecting surface; 5. the spread of a cultural characteristic.*

I never thought to encounter something like poetry in a dictionary. I read the compiler's take on my word and congratulated myself on my choice of noun. Then I thought better of it and instead I congratulated the compiler and the English language on its capacity to invent such a word with so many strings to its elegant bow.

Diffusion. Nice job.

I drifted into a kind of deep sea reverie of surfacing whales with phosphorescence breaking about their flanks, and falling in dripping sheets from the long upraised pectoral fins (they were humpbacks I fantasised, they always are), curtains of folded points of fire. For a few moments the walls of the room faded and the sea had made of me a wholly consenting prisoner. And then in the real sea world beyond the real window in the real wall in front of me, my attention was drawn back to where there was a dark curve and another, and another, and another…the dolphins had come back. I was watching dolphins from the sofa.

Mostly, you work hard to come close to nature, hours of travel and patience and stillness (that above all things) to win a handful of moments' reward. That is as it should be. Your species and mine has engendered in nature a profound mistrust of our doings in wild places, such has been our wholesale and increasingly aggressive abuse of our place in the world for centuries. But if you are inclined to come close to nature, to win a degree of intimacy there so that nature confides some of her secrets in your presence, then that mistrust must be overcome by a dedicated deference and by setting out to encounter nature *on nature's terms*. And as Nan Shepherd put it in her unsurpassed (and probably unsurpassable) little masterpiece, *The Living Mountain* (Aberdeen University Press, 1977): "This is not done easily nor in an hour. It is a tale too slow for the impatience of our age." But if you listen to the land, listen to what nature is trying to tell you, then once in a while nature will reach out to you when you least expect it, and sometimes the results are extraordinary. These are the blue moon moments. Watching dolphins from the sofa was one such.

Of course, then there was the undignified scramble for binoculars, the rush to the door and across the street to the edge of the sea, there to stand and stare at the quiet miracle that is the unhurried grace and ease of a pod of dolphins with somewhere to go. The swimming is matter-of-fact, flirting in equal measure with surface and underwater, devouring distance. It is quite different from the playing-to-the-gallery high jinks they deploy when they ride the bow wave of a ferry or one of those specialist dolphin boats, such as the one I was briefly permitted to steer out in the Firth of Tay a couple of summers ago. When we came back into harbour at Broughty Ferry I thought briefly that nothing could be finer than to skipper a boat like that. But after a while, once I had re-lived the process a few times, once the adrenaline rush of the moment had dissolved, I thought better of the notion, I decided I would prefer to be a dolphin.

But it was Burnmouth that redefined the dolphin in my mind, when nature reached out a hand to me, and I reached out in return and took it and held on. I choose to believe that nature's purpose at that moment was to fasten into place that connection established earlier in the day, a connection that had been growing and strengthening throughout the week, so that it was secure for all time, so that it honoured in the most remarkable way a sense of time and place in nature's relentless pageant. For the umpteenth time in my nature writing life I summoned those words of the English writer Margiad Evans:

There is no substitute, even in divine imagination, for the touch of the moment, the touch of the daylight on the dream.

◉ ◉ ◉

The tide hauls itself away from slack water just beyond the cottage door. The seals feel the change far out in their deep-sea wherevers, respond to its summons, and home in on their low-tide couches, those narrow parallels of grey rock, straight as harbour walls. From the top of the headland above the Apache's head-dress, looking north up the coast at the full extent of those sprawling parallel rows of rock, the effect is of a bird's eye-view of a ploughed field, albeit it one that was ploughed long enough ago for the furrows to have fallen in here and there. To the east is the sea. To the west is the single street of Burnmouth at the base of its cliff. The rest of the world either lives beyond the sea or beyond the clifftop. "The rest of the world" includes the village that the road maps and the road signs proclaim as Upper Burnmouth, and it is as utterly of that land of fields and woods as Lower Burnmouth is utterly of the sea and the cliffs. I sensed that here was a landscape with a story to tell. I looked back to the seals. The reason the seals home in on the place is surely the nature of the place itself. The seaward edge of the low-tide rocks is a perfect refuge to while away the low-tide hours. Eventually it will be raised just above sea level and because it is long and straight and level it can accommodate all the hauled-out seals that linger along this corner of the coast, and offer them a fast and simultaneous retreat into the sea if danger threatens from the land.

I wondered how long the ancestors of this group of seals were accustomed to hauling out here, and if today's generation had inherited any awareness of the pedigree of the rocks. What I looked down on from the top of the headland was the aftermath of nothing less than the original collision between Scotland and England. This coast from the Border north through Burnmouth and as far north as the edge of Eyemouth is officially the Burnmouth Site of Special Scientific Interest, although in truth, the interest extends a little further than the limits of geological science.

There was a time when Scotland and England adhered to two quite separate continents with an ocean between them. This much I knew. Bear in mind that I am writing this while Britain invites the ridicule of all Europe and much of the civilised world for indulging in the orgy of self-harm that is Brexit, and bear in mind too that there are a couple of million or so Scots who have begun to look kindly on that particular geological epoch and to think of it as the good old days. And if you were to tell those same people that a fault runs through the site from south to north they might mutter "tell me about it", so best to make it clear that the fault is also geological rather than geopolitical, and that what it divides is only geological eras. But that "only" belittles the scale of what happened here. To the west of the fault, the rocks are Silurian and somewhere between 420 million and 430 million years old; to the east the rocks are Carboniferous, a youthful 335 million to 355 million years old. Non-geologists might struggle to grasp the significance of this. I only mention it because I am one such and I did – struggle.

So, deep breath, and here goes: the two continents started to drift towards each other. First of all, what causes a continent to "drift" and why are two drifting at the same time? Towards each other? You just know there had to be consequences. One consequence was that the ocean between them was squeezed. Another was that the ocean floor was pushed *underneath* Scotland. Oh, wouldn't you love to have seen that happening from a safe vantage point, just to see how it works? And I know this is basic stuff for geologists, but bear in mind that I, non-geologist that I am, had just found this out and here I was sitting on top of the Apache's head-dress on a beautiful midsummer morning, scrutinising what even to my eye was an unusually formal looking arrangement of rocks, and I had only come here in the first place because of the seals. And suddenly, I was having to contend with the question: how does an ocean floor get pushed? What manner of forces does it take to achieve that, having first squeezed the ocean itself?

Now for the bit you might begin to understand and then recognise. While it was still joined to its old continent, perhaps an outpost of western Denmark or southern Norway, sediments from Scotland's rivers poured onto the ocean floor to form "fans", but as the ocean floor slid beneath the Scottish landmass, those sediments were scraped off, and instead of lying on the ocean floor they were heaped up into substantial mountains. But then, as the geologists might say, "erosion dominated over deposition", and the mountains wore down, and the result is what twenty-first-century Scots call the Southern Uplands.

Then Scotland and England collided. And again, wouldn't you have loved to be there and witness that first forging of two restless neighbours that would become one restless land? Unfortunately, as always with geology, it was just before our time.

The new Scotland's new mountains poured new sediments from its newly reconfigured rivers, and these flowed into what is now south-east Scotland and formed a coastal plain, and finally, this corner of Scotland began to look like itself.

Finally (finally for the moment, for in reality there is no end to the story of the land and its uneasy neighbour, the ocean), when grazing animals finally got here, they found the sea cliffs too steep for them, and a counter-conspiracy of coastal grassland and landslips on the cliffs created a mosaic of vegetation and bare soils ripe for colonising by adventurous plants. Among these is kidney-vetch, a modest yellow cluster-flowered member of the pea family.

All of which explains why, on a warm June morning of 2019 this Scottish nature writer crouched by one such yellow cluster admiring not the flower itself but the small blue butterfly, which rather dotes on kidney vetch. The small blue is a bit of a watercolour butterfly, and it helps if you are looking at the underside of the wings where pale grey predominates and graduates with exquisite subtlety into sky blue. The cliffs are also something of a stronghold for the northern brown argus, which is as boldly handsome as the small blue is demurely pretty. Chocolate brown with orange and white accessories, animated by full-on June sunlight, you never tire of the male's fast flight around its territory. If you do find one on patrol, remember where you found it, because the chances are that you might find it there again when you come back. And be sceptical when your field guide preaches at you that they are difficult to approach. Sometimes, if you are willing to sit dead still, they approach you. Every time I come across this kind of thoughtless field-guide speak I cite the example of the Scotch argus in Glen Orchy a few summers ago, a moment of startling and unfettered joy in nature's company, which, after a few retellings, resulted in this:

Another of the Reasons Why
I'm Suspicious of Field Guides

Scotch Argus (a butterfly not a drink)
lives in Glen Orchy, or more accurately,
I never saw one anywhere else.
Whenever I'm in Glen Orchy (damp,
peaty, sweet-watered, flower-drenched,
Argyll in a nutshell)
and it's the butterfly time,
I expect to see them.

A field guide I own intones:
"*...restless and difficult to approach...*"
yet they keep bumping into me. And here,
where I sit half drunk on bog-myrtled air,
are five of them, one on every fingertip
of my left hand. All of which offers up
another of the reasons why
I'm suspicious of field guides.

With the butterflies drifting a few feet away across the cliff face rich in its summer gladrags of flowers and grasses and shrubs, I was permitted to bear witness to a moment that concluded a circle of events that had taken hundreds of millions of years to arrive here. It is no small thing, such a moment.

And if I might return for a moment to the notion voiced by John Muir – "When we try to pick out anything by itself, we find it hitched to everything else in the universe" – one of the things I like about my day job is how moments echo each other across time and landscape. And thanks to the butterflies at Burnmouth that emerged from a 350 million-years-old geology lesson, I forged a link within my own idea of my own country between its Lowland east and its Highland west, Berwickshire with Argyll.

And then there was a still more startling echo.

A few days after I returned from Burnmouth, I was driving back towards Stirling from Flanders Moss, so as landlocked as mainland Scotland permits. It was a still, early evening, the fields of the Carse of Stirling quiet under their arc of mountains. I was alone, I had my window down, the car CD player was playing Yo-Yo Ma's newly released reinterpretation of the Bach Cello Suites, which to my mind is a perfect pinnacle in the art of making music. Then I saw a red kite hunting low over a field.

The road was deserted. I stopped the car and reached for the binoculars that live behind the passenger seat; I cut the engine, and with it, the music. The car flooded with the warmth of that late-June evening and a profound quiet, and the two fused into a stillness of rare quality. The red kite in hunting mode is both slow and silent, often stopping on the air, head down, fork-tail splayed, sunset-shaded. In this manner it drifted sideways across the field, then back again, a pattern of shallow zigs and zags as elegant as it was purposeful. Then it changed tack and came towards me, so that now I saw it head-on and slowly filling more and more of the binoculars. It was no more than thirty yards away, not ten feet off the ground, when it stopped, held still, raised half-folded wings high above its head, pulled both legs and feet together and stretched them vertically below its body, and in that attitude it fell, and the soft thud of its landing was the only thing I could hear. But the point is this: in the instant before it began to descend, wings high, legs and feet together so that the realigned shape of the bird was widest at the top where its half-shut wings created a pair of matching wedges, and narrowest at the bottom where the closed legs and feet brought the whole thing to an elegant point. What came in to my mind in an instant in that Stirlingshire field was the perfect dive of the Burnmouth dolphin, for the shape of the kite echoed the shape of the other.

Sometimes I think that summer is my least favourite season. Then something like that happens, and what Muir said of everything being hitched to everything else in the universe…it is every bit as true of the seasons, and the reality is that just as there is really only one ocean and we have given different bits of it different names, so there is only one season and we have given different bits of it different names too, and there is just as much chance of glimpsing eternity in midsummer. And when you do, you will find that that glimpse is hitched to everything else in the universe of the seasons. The red kite rose and flew off (it missed), resumed its zigzag pattern away across the field, passed some big trees and vanished into the next field.

Meanwhile, back in Burnmouth, where a butterfly had

crossed a cliff face and my nature writer's mind had slipped its moorings, I was still left with one unanswered question. It nagged me all that week. It nags at me still. What happened to the ocean that got squeezed? And then a second question as unanswerable as the first: is that ocean bed still underneath Scotland, quietly awaiting the advent of one more geological upheaval, perhaps to be rescued from subsea ignominy and become mountain, its new rivers glittering in unaccustomed sunlight? In the process, will we regain our long lost land bridge to continental Europe so that the people of that redesigned Scotland can reclaim our place as a European nation, the one we voted for a few hundred million years ago. That assumes, of course, that there will be someone left to vote.

The wolves would be back, though. All we lack now for wolves to reintroduce themselves is a land bridge, for as I write this they are in every country in mainland Europe again, and no one has been eaten alive because of it. In the meantime, we have the kidney vetch, the small blue and the northern brown argus as a legacy from the last time we heaped up new mountains hereabouts, that and an unsung coast that has found its voice, and I have heard it sing.

Chapter Seven

Between a Rock and a Soft Place

THE NATURE OF THE Burnmouth cliff face is at its midsummer jungliest where you try to thread a waymarked route up from the harbour to the clifftop path that swings north to Eyemouth. Nature does its best to convince you that the waymarkers are lying, or at least they are in on some perverse species of practical joke. I had long believed that brambles are no good to man nor beast until and unless they are clustered with berries in their glossy black edible prime, and midsummer on the east coast of Scotland is way too early for that happy day. I was about to be proved wrong. By a whitethroat. And then I was proved wrong by another whitethroat, and another and another.

But first, once the jungliest infernalisms (sometimes the English language misses a trick and you have to reach for your own personal O.E.D. to explain yourself) of the so-called cliff path and its waymarkers with their out-of-kilter sense of humour (I still bear the scars, or at least the scratches, the stings) have had their fill and deposited you up there in the overworld where the rest of the people live, you might stop and turn around and stare at the suddenly endless scope of the land, and realise that there is still a west out there as well as an east and a north and, to a certain extent, a south, which is all that Lower Burnmouth permits. It is a couple of days since I saw anything that looked like west. I would imagine that spending the winter solstice there, stocked up for Christmas and the New Year, the compass of your life might cheerfully do without west for a few weeks. But back in the here and now of midsummer, somewhere back down there while I was scaling the cliff face, I must have taken a couple of paces across a gulf about a hundred million years wide, from the Carboniferous to the Silurian, and never even noticed. How can that be? How can geology sometimes be so coy about these things?

Up in the overworld, a barley field, of all things, wafted a warm wind towards a thread-wide stretch of the Berwickshire Coastal Path, where it clung to the clifftop and where the jungliest infernalisms peered up and over the edge from below to see what lay beyond, but baulked at the looming frontier of barley and ducked down again out of sight. All but for one of the brambliest manifestations of all, an impudently brandished bare and apparently dead tentacle that contrived to pose against a cloudless sky. It may have been leafless but it still bristled with thorns and leaned up at a rakish angle, for all its apparent lifelessness. And it was there, most of the way towards the branch's ragged-edged and broken-off end, that a whitethroat had a firm enough grip on a space between two thorns (but perilously close to both of them) to be untroubled by the effect of the breeze on the branch. It rode the branches and sang at the same time, and even as the breeze ruffled the pale brown feathers at the edge of its song-throbbing throat (a bib of brightest of white in the sunlight: this is a bird that does what it says on the tin), the volume and the sweetness and the inventiveness of the song exceeded every expectation. We, the whitethroat and I, were so aligned that I could see right into the pink-walled inner sanctum of its tiny singing mouth. Somehow, that made me uneasy about the binoculars, about the camera with the zoom lens. The bird was not twenty feet away. I felt oddly intrusive, my sophisticated optical equipment suddenly excessive, two sledgehammers to crack a very small nut. Yet the bird behaved as if it were quite unintruded upon, as indifferent to my presence as the sprawl of Silurian rock somewhere beneath my feet.

As if it read my mind, the whitethroat stopped singing, leaned forward and stared surprisingly fiercely and directly into the lens of my pointing camera. It's not what you expect from a four-inches-long speck of grey-brown anonymity on a dead bramble branch. Its beak snapped shut, and in its new attitude the whitethroat fell into shadow and became grey, robbing the bird of its one distinguishing feature, yet it still stared and still achieved a sense of presence I found slightly unnerving. I put it down to the bird's fearlessness. I am six feet tall, and it's four inches from nose to tail, but there was no mistaking the air of challenge in the stare. And look at where it was perched: in the glasses I could see that its feet rested *against* the sides of the triangular bases of two brutally erect thorns. This was a bird that knew about the art of perching with millimetric precision, about living dangerously. Then it flew surprisingly close to my right shoulder and disappeared behind my back and over the clifftop.

A cursory look over the edge revealed where it had gone, where it nested, as it turned out: a small and improbably steep plant community composed entirely of brambles and nettles. Another of the bird's skills, then, was an acute fluency of hopping flight *inside* bramble bushes. So not only is its nest site protected against pretty well everything this side of a wildfire, there is also a source of doorstep food just before it migrates at the end of summer or the beginning of autumn: ripe brambles, the perfect fuel for lion-hearted whitethroats hell-bent on a flight that goes south-east to France, then south-west to Spain and Portugal, then south to Africa. Curiously, when it returns in the spring, it cuts out the dog-leg and just flies north. There will be a reason. I just don't know what it is. But what I do know is that the whitethroat is a sophisticated little bird that lent its sophisticated voice to embellish my unsung shore with one more variation of song.

◉ ◉ ◉

The clifftop path lies between a rock and a soft place, a clifftop and a barley field. The midsummer barley field sways to the west and back again in response to the breeze, a honeyed shading on the face of the land, rising and falling like breathing. The barley presents a neatly cut straight edge towards the sea; a golden wall, but a wall that whispers, an agreeable restlessness. Because the path lies at a slightly lower level than the field, the whiskered tops of the barley stalks rise almost to eye-level. And because I had climbed the jungly cliff-face from Burnmouth's sea-thirled front door, the first time I saw the waving top surface of the field it put me in mind of the sea at Lindisfarne, where the shore lies below sea level and the sea sways shorewards from a gently crumpled horizon. The obvious difference is that the field has that neat, straight, tall edge, and it is that that replaces the advancing wave of the sea, and that when the wave breaks it curves towards your feet, not your eye-level, and sometimes it stops half a mile away. Always, where the land meets the sea all along my native shore, there are two frontiers, one at the bottom of the cliff and one at the top.

The edge of the field, between the path and that frontier of growing barley, is the skylark's place. Watch it rise from the tussocks on a silent whirr of wings, and within five seconds and ten vertical feet, the singer begins. Wherever I have walked along the edge of my native shore between Tay and Tweed, and wherever fields sprawl towards the sea, there is a benevolent brambled and tussocked no man's land, and if the two are close enough together, and if you are willing to sit still and listen for a space in your day, you will have earned the right to hear whitethroat and skylark overlap in song.

Compared to the skylark, the whitethroat is a miniaturist, dealing in flourishes of a dozen or so seconds at a time, but when it is in the mood, the flourishes go on and on. The skylark is a climber and sings from floor to ceiling, and the longest I have ever timed one is seven minutes of unbroken song, and even in descent it sings until once again it has dropped to somewhere around ten feet – and there the song cuts out and the silence that follows is the emptiest you will ever feel on that coast. Unless, that is, there is another skylark a little further up or deeper into the field and the distance lends it the character of an echo. If not, then a whitethroat steps onto a perch and the air is all diamonds of song again. And then the rhythm section kicks in when you least expect it. Three notes, percussive and piercing, not loud, but then a good rhythm section always defers to singer and soloist. Again and again the three notes, and if you wait where you sat to make space for the skylark and the whitethroat, one or other will pick up the song and run with it. Three or four times over the years I have heard the three coincide – skylark, whitethroat and reed bunting (for that is the identity of the rhythm section). And because I have a bit of a musician's ear that I can bring to such moments, I have wondered if there was a common ancestor to both small group jazz and chamber music, and if so, was this it?

◉ ◉ ◉

My first impression of the seabird cliffs of the red sandstone Berwickshire cliffs had been a raw profusion of life up at St Abb's Head. They had been crammed with birds from clifftop (puffins up there miscast as sentries at attention by their nest burrows, or flying in with their preposterous beaks stuffed with sandeels) to high-water mark (kittiwakes down there, sleek beauties of the surf), with razorbills and guillemots on the impossible ledges, fulmars slotted in anywhere and everywhere. More birds crowded the caves and stacks, they crowded the sea and they crowded the airspace, the rock in all its forms and the surface of the sea, and heaven knows what formidable divers thronged the underwater underworld and how they deep they penetrated the gloom. It seemed like the only thing that might challenge the bird count was the decibel count. If anything detracted from the bird spectacle itself, it was the din.

All that was a midsummer day not unlike this one, sometime in the early 1980s. It did not feel like a long time ago. But walking today's clifftop path north from Burnmouth towards Eyemouth and St Abb's Head beyond, that other day thirty-something years before drifted back into mind, and what it put there was the memory of that din, that cacophony, that raucous chorale of raised voices, and what put the memory there in the first place was the complete absence of din. The thought took root: the glass-half-full-glass-half-empty analogy wouldn't work there. There was nothing half full about it. These cliffs, that sea, that airspace were almost all half empty. The ones that weren't half empty were just plain empty. I was dragged away from the thought by the voices of kittiwakes drifting up from a stack in an all-but-silent sea. I peered over the edge with the binoculars and found them. There were six. Six!

Kittiwakes have been one of my symbols of summer forever. I have already explained that I learned the word when I was very small, because on trips out from Dundee to the coast at Arbroath or Auchmithie I had heard the bird call out its own name. I had an eye and an ear for such things from the start. But a bird that spoke its own name and lived where the water was at its stormiest, its whitest, its noisiest, endeared itself to me effortlessly.

The emblematic evolution of kittiwakes through all the summers of my life, from earliest childhood to seasoned nature writer, experienced something of an apotheosis out on one of the northmost extremities of the land: Eshaness on the Shetland island the natives call "Mainland", and about ten years after I first walked this Berwickshire coast. That particular Shetland day, one of the most impressive storms I have ever seen hurtled across the Atlantic and hurled itself at the long Eshaness coast with its countless stacks and skerries, arches and caves and riven cliffs. The names tumble over each other in my mind even now, for I had never heard their like: The Grind of the Navir, Scraada, Moo Stack (mysteriously, there are two of these about seven miles apart), The Bruddans, Utstabi, The Runk, and my favourite – The Drongs. The sea charged 100-feet cliffs and heaved itself clear over the top to dump salt water in the peaty freshwater lochans beyond. I watched from the lee of a rock, thrilled beyond words, stupefied by the thunderous beauty of nature. It was a partial shelter, at best; there was no real shelter, not there, not then, but I was as cheerfully contented with my nature-writing lot as I have ever been. Not least because there was barely a bird in the sky that wasn't a kittiwake, but there were – oh, I don't know – hundreds of them. And they hung in a fragmented cloud that rode the ocean-driven wind far higher than the clifftops at its zenith and thrillingly close to the breaking surf at its most daring. It lazed and drifted and its bird-fragments changed places up and down the gusts and the thermals and the sun shone fitfully among them from broken clouds that perfectly matched the broken sea and the broken coast, and for a moment the birds were bejewelled. And all the while they yelled out their name so that all the wild world would be in no doubt who dares to fly in the face of the storm. Sitting by that rock, half-deafened by the pile-driving ocean, I fancied I could remember the child that first wrapped his tongue round that name. And a few years before Shetland, there had been St Kilda, and The Tunnel.

◉ ◉ ◉

From there it had come down to this, six kittiwake voices at the base of a cliff above a flat calm sea at what should be the height of the nesting season. One cliff face after another told the same story. The quiet was eerie. I had half-expected something like this. After all, the decline of seabird populations around our coasts and around the northern hemisphere has been well documented. Biologists have pronounced on it so often that perhaps we grow a little deaf to its message. But in addition to being a full-blown natural disaster in its own right, and with serious overtones of unnatural disaster thanks to humanity's role in the seabirds' plight (our fingerprints are all over it, from plastic-strewn oceans to overfishing, from melting icecaps and glaciers to interference with the movements of fish shoals), it is a self-evident symptom of the wider malaise that is our chaotically out-of-kilter climate.

The bird-dearth struck a painful, jarring discord on such a beautiful day on the Berwickshire Coastal Path. At the top of the cliff there were the considerable consolations of the sea view, the three-fold chamber music of skylark, whitethroat and reed bunting, the sudden splashes of poppy and ox-eye daisy, the infinite sky. But between clifftop and cliff bottom, that vacuum once filled with seabird clamour hung like a lead weight on the day, on the land, on the sea, a lead weight that hung from nature's neck and dragged it down.

Chapter Eight

City of Ghost Birds

THAT STRETCH OF COAST PATH north from Burnmouth reached a natural denouement on a headland above the little seafaring town of Eyemouth, which clusters around a harbour that still works for a living. It looked trig and trim from high above, but even as I paused to admire the town's quiet self-containment and the wider view of gathering land and sprawling sea, my eyes were drawn to a newly revealed swathe of coast, for at the heart of that wider view was the unmistakable deep-red, guano-drenched cliffscape of St Abb's Head.

The place is a byword among ornithologists and coast-path walkers. Its reputation is for seabirds and geological wonders in complementary spectacle, for wildflowers, clifftop lighthouse, and for wildfowl and wetland at the Mire Loch behind and inland from the cliffs, all of it threaded with footpaths. But most people come in spring and summer and most people come for the seabirds. From that headland above Eyemouth the apparent profusion of guano was a welcome sight, a pale treasure-store of hope, like a glimpse of the North Star to a fogbound seaman. I never thought I would be so pleased to see bird shit. But I was about to find out that what looks from a distance like bird shit in profusion does not necessarily equate at close quarters to birds in profusion. But at that moment it suggested at least that St Abb's was faring far better than the cliffs where I stood.

I first got to know St Abb's a bit in the 1980s when I worked as a newspaper journalist in Edinburgh, and spent some of my spare time exploring the coasts of East Lothian and Berwickshire, from Aberlady to the Border. First impressions were defined by non-stop flypasts of slow-wheeling fulmars stalling on the wind and hanging in front of my face on chittering wings to stare me out, uplifting choruses of kittiwake voices, and guillemots, razorbills, terns, puffins, with peregrines to keep them all on their toes and drive the mass screaming into the air; and the long-haul wave-top safaris of gannets; the symphony of it all and the stench.

To this day, and now more than ever before, a species of walk-this-way-tourism I don't much care for has thoughtlessly, carelessly branded St Abb's Head as "seabird city" (anything less like a city…). It is the kind of advertising-speak that is not just thoughtless and careless but also dangerous, and it is everywhere in the promotional material of sundry internet companies that pitches to the lowest common denominator, with its reckless use of "stunning" and "iconic" and which thinks that the only yardstick for success is more and more people spending more and more money, and who cares what has happened to the kittiwakes? (A seventy-eight per cent decline in less than twenty years, since you asked.) Tourism is a crude master for places like St Abb's Head, and nature recoils from the onslaught. Even the National Trust for Scotland is culpable. As owners and managers of what is a national nature reserve you would like to think it would know better, but its publicity material is designed to lure ever more visitors with variations on the "seabird city" theme (it also uses those very words), and then, apparently without irony, it gently rebukes those same visitors following its own research which found that "breeding success for both guillemot and kittiwakes was significantly reduced by the presence of people because the nesting birds experienced stress, causing raised heart rates. This resulted in additional energetic cost to the birds that could lead to nest desertion..." Did it really take scientific research to conclude that nesting seabirds don't like the constant proximity of lots of people throughout the day? And could it be that a new visitor centre and much larger car park have anything to do with it?

I chose a quiet midweek evening to see whether the guano on the cliffs justified that stab of optimism back on the cliffs above Eyemouth, or whether it signified nothing more than the lingering stains of old years of plenty. Or had nesting birds turned up in numbers, found a dearth of food in the sea and left again without breeding? The first bay I came to was all but silent, its airspace and its cliff ledges devoid of everything the "seabird city" language implies. Those first few moments set the tone. The walk soon deteriorated into a search for straws to clutch. I found some, but none of them involved seabirds, for the meagre handfuls were as eerie as ghost birds. The evidence of absence was barely believable, except that it was all too visible. And where were the puffins? In all the places I knew where to look for them, I could find none at all.

So I pulled out. I opted for the Mire Loch and its mute swans. So often in my nature-writing life I fall back on swans. From ubiquitous mutes like these to Icelandic whoopers and the compact Siberian Bewicks and (once, in Alaska twenty years ago) the mighty trumpeter swans, with a wingspan nine feet long and as wide as sails and a voice like Louis Armstrong's horn. Aah, you had to be there, that summer when I was introduced at very close quarters to grizzly bears and humpback whales and bald eagles and unclimbed mountains with no names and a Tlingit native who called me a white man to my face then told me that "it wasn't God that made the world, it was Raven, and his first job was to make the world in perfect balance and he made the bald eagle with a white head and tail and black body, and that was the symbol". All that coruscated through my mind in a moment, and with the cliffs behind me and the swans ahead of me I thought to myself:

"So what the hell happened to the world in perfect balance?"

Meanwhile, the mute swan pair and their three cygnets cruised through the ducks, coots, grebes and lesser fowl. We become inured to mute swans, take them for granted, perhaps because of their ubiquity and their willingness to share our redesigned world. But their lifestyles are as extraordinary and diverse as the habitats where you find them, from city canal to wild mountainside loch and wilder island loch. My favourite mute swan nest is at Stenness in Orkney with a view across the sound to Hoy. I have made something of a study of the swan tribes and written about them more often and more thoughtfully than anything else in nature, and the sight and sound of them in flight moves me like few other creatures.

On the Mire Loch, their presence constituted one of those straws I clutched. It's not that they assuaged the gloom engendered by the crisis on the far side of the cliffs, but their individuality and adaptability have made them one of nature's great survivors, and sometimes when humankind pushes the specialist tribes like seabirds to the far edge of their comfort zone and beyond, there is respite of a kind for a troubled nature writer to pause and take stock in a theatre of nature where all is well. The landward side of the cliffs above the loch is as smothered in profusion as the sea-facing cliffs are devoid of it, an extraordinary transition is effected, and my mood transformed at the sight of this other world that lies 100 yards in a straight line from the ghost-tormented seabird cliffs just out of sight. Here the slopes are smothered in flowers: orchids, sea pinks, sea campion, bedstraw, stonecrop, red valerian, birdsfoot-trefoil, kidney vetch, thyme. The widespread juxtaposition of thyme and birdsfoot-trefoil – that deep purple and bright yellow combination again – is a sumptuous patchwork of nature as daring colourist, and it works, it works! Two stone walls nearby offered up a smooth hawksbeard and a perfectly erect shrub-shaped honeysuckle, both of them rooted and finding sustenance between stones on the top courses of the walls. Sometimes nature makes you scratch your head and smile.

But walking back on an inland footpath through fields and a landscape of plenty (the Silurian coastal plain, not the Carboniferous coastal frontier), the spectres of the seabird cliffs crawled out of their hiding places, spreading their unease like melting ice sheets. I struggled with the unease of that troubling co-existence of famine and feast. You could wander this side of the cliffs and smell the honeyed air and bask in a warmth the onshore wind would deny you out on the shore, admire the colour show and the honeysuckle casting its magic spell about the labours of some long-gone drystone dyker (and what a glorious endorsement of that day's labour, decades ago), and you could linger by the flag irises on the shore and trace an otter's path there, watch the swans impose their exquisite regime on the watersheet and its shores, lift your eyes to admire the peregrine's flat-out annihilation of the one white pigeon in a passing flock. You could conclude then that you had spent two of the most agreeable hours in nature's company, and never raise your head above the parapet that lies above you and between you and the seabirds' doomed domain.

But the crisis on the seaward side of the parapet afflicts more than a few miles of cliffs between St Abb's Head and Burnmouth. The entire thousands-of-miles-long habitat of the coast of Britain, from Land's End to John o' Groats and from Ardnamurchan Point to Ness Point at Lowestoft, and all its islands from Wight to Unst and St Kilda to Farne... and beyond all that to embrace the reach of all the world's oceans: that is what is dying here. Tourism blithely – naïvely – trumpets the second highest cliffs on the east coast, yet these cliffs constitute a tiny symptom of that immense malaise. And just how critical is the health of that tiny symptom?

In 2018, Scottish Natural Heritage published the results of a survey of the seabird population between St Abb's Head and Fast Castle just up the coast. The survey covered the years 2000–2015 and comparisons were drawn with previous surveys in 1987 and 2000. The latest survey is particularly timely because there is a new factor about to come into play: the construction of major offshore windfarms off the coast of south-east Scotland. So would you like the good news first? Cormorants have increased by twelve per cent since 1987.

As I started to read down the list, the same question that had been at the back of my mind since I started to walk the cliff path on that June day in 2019 now returned to front of my mind: where were all the puffins? I got the answer, but first, apart from the puffins, the rest of the list looks like this:

Fulmar: seventy-two per cent decline since 2000.

Shag: forty-seven per cent decline since 2000; seventy-three per cent decline since 1987.

Herring gull: seventy-eight per cent decline since 2000.

Kittiwake: seventy-eight per cent decline since 2000.

Razorbill: thirty-five per cent decline since 2000.

The study added a puzzling and unexplained footnote. The rates of decline at St Abb's Head to Fastcastle are "more severe" than the national picture, and the rates of increase are "far less". And that was before the first of the new wind farms is built.

And then there was the answer to my puffin question, a nightmare in a statistic.

Puffin: 100 per cent decline since 1987. Which is another way of saying that the total number of puffins is none, nil, nothing, zero.

Gone.

◉ ◉ ◉

To sing a song for an unsung shore: it is a peculiar quest. It is not as if I imagined a shore that had no voice, for all landscapes sing. More often than not their songs are unheard and disregarded, but only because of our unwillingness to listen.

Listening to the land has become my life's work. It is how everything I write begins. Now here, on a tract of my native east coast was a land I had never listened to before, and as far as I know, the literature of the nature of Scotland has never paused here for any length of time. Something seemed to fall into place when I found the cottage at Burnmouth, and something else fell into place when, having booked it for a week, I eased my car gently down the steep and narrow brae that defends it from the rest of the world above and beyond, negotiated its single-sided village street, parked on the shore opposite the cottage, stepped out and heard the seals. I told myself:

"Listen."

It's all I ever tell myself when I meet a landscape in full voice. Just listen.

All that week, summer blessed the enterprise. The east is Scotland's sunniest coast by far and all week it rained sunlight. Sunlight lay on the surface of the land like glossy floodwater. The land bathed in it. As it was the second half of June there was nothing that approximated to darkness. Flowers and butterflies began to swarm. An invasion of painted ladies began that would make national news headlines and fill the letters pages of those national newspapers that care about such things. Small birds girded their loins for second broods and birdsong enjoyed its second spring, grace notes to underlying seal song. The cliff face immediately behind the cottage was terraced into a garden that accommodated a decking area with outdoor sofa and coffee table that looked over the roof to the sea. From there (assisted variously and at different hours of daylight and endless dusks by pork pies and whisky, tea and coffee and cake, cheese and wine), the song I sought began to find its voice. With a notebook in my lap, pen in hand, cup in hand, glass in hand, sometimes just a contemplative pose of hand in hand, I began to hear the sense of it, and something beyond the physical landscape was caught and held. Right time, right place. Predestined.

The seals were the constant element, the song's recurring theme arose every low tide, whether mid-morning, mid-afternoon or midnight, their wolfish cadences as fundamental as *ùrlar* to pibroch – the ground, the bedrock.

I have written elsewhere in this tetralogy of the seasons that I like edges, edgy places: in particular, the edge of the Highlands, the edge of the Lowlands, where the two collide, the brief landscape fusion where they overlap. Likewise the edge of the land fused with the edge of the sea satisfies a variation on the theme, the sense that whatever direction you follow when you cross that edge, beyond all is changed. The creatures of land and sea overlap, too. Seals haul out for low-tide hours and in season their pups crawl up the beach; kestrels haunt the low-tide edges and foxes wade the shallows, and a ringed plover might drag a wing to lure one away from its nesting ground. Some foxes fall for it; some know better. Later, they will lick the salt from their fur, the sea savour from their land clothes.

The east coast rarely hits you between the eyes, even in summer, not like the west coast of Mull or the mountains of Knoydart or Torridon or Sutherland. It creeps up on you when you least expect it, pale turquoise sea at 3a.m., the slow motion kaleidoscope of the sea light you only tune in to slowly because the sea light insists on it, the chamber music of three small and inconspicuous birds, dolphins materialising out of a flat sea and you never saw or heard them coming, seals stirring into haunting song out of long silences with no obvious reason for either song or silence. The bit that used to hit you between the eyes is – was – the seabirds. Roy Dennis, ornithologist of rare distinction, once referred to Scotland's east-coast seabird colonies as Scotland's Serengeti, a reference to the sheer profusion of animal life. It was a perfect analogy. This unsung shore epitomises the calamity that unfolds before our eyes. We have tampered and we go on tampering with the ocean and its creatures too often and for too long, and the Serengeti of our shores drifts apparently unstoppably towards desert.

Puffin: 100 per cent decline since 1987. That one hit me between the eyes.

◉ ◉ ◉

A strange thought occurred to me:

"If the puffin becomes extinct, they will never forgive us."

It was the form of my own words that I found strange. Who are "we", the committers of the crime, and who are "they", the unforgiving ones?

It may seem like a quite ridiculous notion (it *is* a quite ridiculous notion) but I keep encountering it everywhere, often in the unlikeliest places, the unlikeliest people. The crime against nature that is global warming is somehow worse when the victims are puffins. Puffins matter to people who can't tell a tern from a tystie, or a gull from a guillemot, and don't care that they can't. They buy puffin cards, puffin tea trays and puffin tea towels, and those mugs where it says "puffin" under an image of a puffin on one side, and on the other side, where there is no image at all, it says "nuffin". Puffins are fun. Puffins are cute. You can make jokes at their expense and they don't mind. They are playful as well as decorative. They don't know their place, and they will not be held in check by the science we call ornithology. They have escaped from inside the covers of the *RSPB Handbook of Scottish Birds* to become the star attractions in bird reserve shops. Puffin soft toys, some of which sound like a puffin when you squeeze them. In a moment of quiet desperation, I googled the words "puffin cartoons" and there were 19,900,000 results, or about 14 million more than the world population of actual puffins. They included "how to draw puffin cartoons", a cartoon of a puffin applying make-up to its beak to attract a mate, another of a young puffin going to its pipe-smoking father for help with its homework and the father puffin says, "Not just now, I'm a puffin", and a trailer for a television cartoon in Northern Ireland called Puffin Rock for three- to five-year-olds. And did I mention puffin wallpaper? Puffin aprons? Puffin salt and pepper? Puffin duvets? You don't get these with shags. It's all a bit too ridiculous for comfort, certainly mine. Yet that chilling absence of puffins from the St Abb's Head cliffs wouldn't let go of me. And that's when I thought:

"If the puffin becomes extinct, they will never forgive us."

And my clumsy thought processes went from there to considering whether or not there can ever be a more telling symbol than the puffin for the head-on, slow-motion collision between the grotesque misuse of natural resources by our species and the wellbeing of the planet. It is, I suppose, at least arguable that more people will pay more attention if the arguments are couched in terms of puffins than, say, white rhinos or mountain gorillas or tigers. Or shags.

The list of threats to puffins is a classic global warming dossier. The world's oceans are a mess and getting messier, littered with billions of tons of plastic, and striped evil colours by oil spills and shipping waste and other pollutants spewed from the people-corrupted land. And outwith the breeding season, the ocean is precisely where puffins live – the North Atlantic and its close cousin the North Sea. The full name for our puffin is the Atlantic puffin. And the North Atlantic, like the rest of the world's oceans, is getting warmer. An upward shift of one degree is enough to move plankton further north; the huge shoals of sand eels follow the plankton, and sand eels are the puffin's preferred food choice. The result is that puffins have to fly further to find food, which costs them dearly in terms of energy, with a consequent drop in body mass. If they can't find enough food, they will feed themselves and not their chicks. And that is the beginning of a one-way journey to extinction. Most of the governments of the world (but not all of them: thank you, President Trump) have signed up to a target of slightly more than a rise of one degree in ocean temperature, which may seem noble and determined of them if they can achieve that, but it is no consolation to the puffin.

What was it John Muir said?

"When we try to pick out anything by itself, we find it hitched to everything else in the universe."

The ocean is also becoming more hostile. More extreme weather events lead to more puffin wrecks. This unfortunately named phenomenon occurs when protracted onshore winds wash up rafts of dead birds; sample tests show that their stomachs are empty. So they have starved first before being hammered into shoreline rocks. Effectively, we are killing them twice. The east coast of Britain is among the most vulnerable wreck sites. One in 2013 was sufficiently grotesque to make national headlines.

Over-fishing adds another intolerable burden. In 2018, a study of all seabird populations by Aberdeen University with counterparts in France and British Columbia compared the impact of fishing between1970–89 and between 1990–2010. Annual bird consumption of fish was down from 70 million to 50 million tonnes while annual fishery catches were up from 59 million to 65 million tonnes. And 2018 was the year that a survey of puffin numbers in Shetland revealed that they had fallen from 33,000 pairs in 2000 to just 570 individuals. The International Union for the Conservation of Nature (IUCN) categorises the Atlantic puffin as "vulnerable" in a 2018 revision of its Red List of species threatened with extinction. The European population of somewhere between 4,770,000 and 5,780,000 pairs is ninety per cent of the global population, and it is projected to decline by fifty to seventy-nine per cent between 2000 and 2065.

The numbers and the threats pile up as fast as the puffin's prospects decline. In Iceland, the world epicentre of the Atlantic puffin, the population has fallen from 7 million to 5 million. The director of the South Iceland Research Centre said in an interview that appeared everywhere from the *New York Times* to the *Independent* that the puffin is the most common bird in Iceland, before adding:

"It is also the most hunted."

Trophy hunters kill thousands every year, as if an apparently irreversible trajectory towards extinction wasn't enough to worry about. It is hard to see the "sporting" element of shooting such a bird, given that, as any casual visitor to the Isle of May or the Farne Islands will tell you, you sometimes have to ask them politely to back off a few yards because they are too close for the camera lens. The Iceland centre has also been tracking puffins for the first time on their ever-lengthening journeys to find food, a research project that is taking them into uncharted territory. One of the scientists involved in the project said that "everywhere, they are going further than we thought".

The director was very clear: an increase in sea temperature is the key environmental factor. He also said this of Iceland's puffin population:

"The millions are deceiving. These birds are long-lived, so you don't just see them plummeting down. In the long run, it is not sustainable."

So I asked myself another question. How long is "long-lived" in a bird as small as a puffin (about ten inches), which lives a life in a singularly hostile environment where it is increasingly difficult to find food? Ten years? Fifteen?

The answer was a bit of a shock: thirty to thirty-five is common and the oldest known bird lived for forty years and ten months before being killed by a raven or a falcon in Norway. So you can see how that kind of lifespan masks the impact of a catastrophe in the making. And if I am honest, I was still clinging hopefully to the fact that there are still millions of puffins out there – somewhere between 12 and 14 million mature individuals, according to the IUCN – when I found an article in the journal *Birdlife International* headed: "Seven birds you won't believe are threatened with extinction." The main photograph is of a puffin. You might be interested in the other six on the *Birdlife* list: turtle dove, snowy owl, yellow breasted bunting, grey parrot, and finally the one that rather slips under the radar, the puffin's near neighbour at St Abb's, the black-legged kittiwake. Just as you don't get duvets patterned with shags, no one makes television programmes for three- to five-year-olds about kittiwakes.

Chapter Nine

Bass Notes

My Burnmouth week had a boat trip built in from the outset. For years I had wanted to sail out to the Bass Rock. Being in the company of 150,000 gannets was only one of the reasons why. The other, the more pressing in my mind, was to examine at close quarters the Bass itself. The Rock.

I really only knew it from the far side of the Forth, from the upper jaw of the Firth, from Fife, specifically that stretch of coast between Fife Ness and St Monans, twelve miles away from the out-on-a-limb headland of the Ness, and still ten miles of open water away from St Monans. From such distances the Bass rather settles into its East Lothian hinterland and the blunt silhouette of the Lammermuir Hills, spired and spiked along their skyline with wind turbines as tightly packed as spruce trees in a spruce plantation. The Bass suffers a bit by association, identifiable more by its pallor than its essentially monumental nature. In midsummer noontides, it wears an indistinct shade of bluish-off-white, the product not just of the backlighting of a south-making sun, but also the restlessness of gannets in epic numbers blurring its contours and thickening its immediate airspace, not to mention the colossal spread of their essential byproduct, also known as guano, also known as bird shit. A favourite pursuit from that stretch of Fife coast was to photograph the Bass with the furthest reach of my camera's zoom lens whenever I could juxtapose it with a solitary passing yacht, or an inshore trawler, a lobster boat, any species of small boat, accompanied by the conceit of quoting my own St Kilda poem to myself:

> We who wave our puny daring
> under Stac an Armin
> creep tinily by.

Then, in the evening, as midsummer suns dawdle and dip down towards the north-west, the Bass explodes with light and stands forward from the Lothian shore, its islandness restored. That low-to-the-horizon brilliance that sizzles across the miles of open water reinvents the rock, picks out every fissure and gully and cave and also illuminates the smooth deep grey places of the north face where even gannets cannot find a toehold, far less a nesting place.

All that I had only seen through a long lens and the wrong side of the river. Now, as I drove towards the Scottish Seabird Centre in North Berwick (perversely, several miles inside East Lothian from the Berwickshire boundary) and veered west at Tantallon's glorious fourteenth-century red sandstone clifftop castle, the Bass materialised in some majesty on its moorings no more than two miles offshore. I understood then that from every angle and in every light and whatever the distance, the Bass at peak gannet season is both irresistible force and immovable object fused into a landmark of nature as great and compelling in its landscape as any mountain.

But its show was about to be stolen. By puffins.

I was still smarting from that sucker punch that hit me between the eyes at St Abb's Head. And besides, no one had told me about Craigleith, an inauspicious little island half a mile offshore, and which turned out to be on the boat's itinerary, too. It is everything the Bass is not: low-lying, inconspicuous, grassy, supporting a vigorous plant community, and where the Bass is virtually a gannet monoculture, Craigleith is biodiverse. As the boat approached, the water was suddenly awash with puffins, and then so was the island. I could have stood up and punched the air, except that I am not a particularly demonstrative creature, and the boat was so well-filled that if I had punched the air, I might have taken out two fellow passengers with my follow-through. And there were also eider ducks, fulmars, grey seals, razorbills, great black-backed gulls, cormorants and shags. Oh, and rabbits. And Craigleith has a bit of a story to tell, for it is at the heart of the Scottish Seabird Centre's SOS Puffin project.

Of all the reasons I have heard for the decline in puffins all across the North Atlantic, the one that accounted for a population crash on Craigleith was in a class of its own: an invasion of tree mallow. This native of mainland Europe's western shores is a favourite garden plant, but given the chance, it can grow in dense thickets up to nine feet tall. Its arrival on Craigleith was disastrous for the island's substantial puffin colony (28,000 pairs in 1999, down to 1,000 in 2007) because it colonised their nesting areas and made impenetrable barriers between bird and nest burrows. The SOS Puffin project recruited a squad of volunteers to attack the tree mallow every October, once the puffins had gone, and slowly, the numbers began to recover.

The mean twist to the story is that the seeds of tree mallow float, and that once they had got into the water around the East Lothian coast, the floating seeds could have been delivered onto the island unwittingly by swimming birds – like puffins, for example. The project goes on, for tree mallow is a tenacious beast, but at least, as I write this, and for the moment, the puffins are ahead of the game.

But there was still the matter of this:

"If the puffin becomes extinct, they will never forgive us."

It had been just a random thought, and I have many such random thoughts in the course of a day and mostly dismiss them. This one rankled. It niggled. It squirmed its way into a different corner of my brain and went to work there. So I stopped and looked at it for a while, and now, what I think is this.

First, you must buy into the proposition that you and I are nature and not a superior creature outwith and above nature's laws. Otherwise, there is simply no hope, all is lost, and it won't be just puffins you will mourn, because sooner rather than later, and most likely sooner, ecosystems will cease to function because we will have ruined all their resources, and without natural resources our species is doomed, just because you didn't buy into that simple proposition. We *are* nature's creatures. Throughout this quartet of books I have advanced the cause of listening to the land. It seems to me, as a nature writer and relentless observer of nature, rather than a scientist, that the puffin-bereft cliffs between St Abb's Head and Fast Castle articulate a clear example of the land talking in a tongue that anyone can understand. The essential message is that if it can happen here – and it just did – it can happen anywhere. "Puffin: 100 per cent decline since 1987." The clipped sentence, its bureaucratic starkness, is among the most chilling I have ever read. What can be worse for a species than 100 per cent decline? It means, "You're too late."

Nature has put the plight of the Atlantic puffin before our eyes for two reasons. One is that what is happening to it is simple to understand: the behaviour of our species is ruining its essential habitat and food supply. The other is that at an unthinking, aesthetic level, we love puffins and we still go to great lengths to watch them just being puffins. Two summers ago, on a visit to the Isle of May four miles off the coast of Fife, I eavesdropped on a small group of fellow-passengers as the boat headed back for the mainland. It became clear that they were regular travellers to puffin colonies all around the coast of Britain. They agreed among themselves that Isle of May was nothing like as good as the Farne Islands off the Northumberland coast, because there you could see more birds and get much closer, and you didn't have to walk so far. (The May that summer had 40,000 pairs.) They talked of nothing but puffins. Much of the island that day had been a shimmering sheet of white, a field of sea campion delectably animated by sunlight and sea breeze that rippled over the contours of the island like a glacier: it generated not a syllable of recognition, nothing. There were nesting eiders right beside the path that eyed you from the depths of one of nature's most perfect stillnesses: nothing. There were waltzing clouds of Arctic terns that flared up from a standing start into the afternoon sun and simply dazzled the beholder's eye: but they were unbeheld by the puffineers. In a slightly troubling way, they were puffin crazy. They are far from alone. And perhaps nature is suggesting that there is something to work with here, that the puffineers are themselves a resource that can be redirected to champion nature's cause through puffins.

Listen to the land.

What it says is this: the Atlantic puffin is doomed.

In its greatest stronghold – Iceland – its natural defences are down. The experts say so, the professional scientists, even while they are still surrounded by 5 million puffins. It is a deception. It is not sustainable. Among those who are not listening to the land in Iceland are hunters, trophy-hunting for puffins. And the hunters, too, are fooled by the deception...look, there are millions of them, plenty for us!

The land is also saying this. Consider the relative demands the two species make on the planet: *Fractercula arctica* and *Homo sapiens*. The Atlantic puffin eats sand eels, and no more than they need. The idea of overfishing is incomprehensible to them. They make burrows in grassy clifftops for their nests. That's it. In winter they head out to sea and look after themselves. What they ask of nature is just enough fish and a clean ocean. That and a climate in which extreme weather events are rare enough to be manageable. They are scarcely more demanding of the planet than fresh air.

On the other hand, the demands our species makes on the planet include this one: we want it all and we want it to do our bidding. One of many consequences is that we deny puffins a reliable supply of its life's essentials, this bird that we profess to love so ardently that year after year, summer after summer, we fill boats and sail to offshore islands to feed our two-faced puffin love. "Two-faced" because at the same time, the species to which we belong is also sealing the puffin's fate. Unless we listen to the land, then, unless we pay heed to what it asks of us, the puffin *will* go extinct. The case is so unambiguous. That random thought that provoked this wee polemic – "If the puffin becomes extinct, they will never forgive us" – is clearer to me now. "Us" means the entire human species wherever we impinge on the North Atlantic. "They" is nothing less than every other tribe of nature with which we share our portion of the planet; it will finally be clear to them that if we reveal ourselves to be incapable of accommodating puffins in our twenty-first-century world order when their demands are so modest and simple, what chance is there for complex ecosystems? And the answer to that is a four-letter word.

Perhaps nature is banking on our love of puffins, that it may yet persuade us to pull back, to remind us that for all our worst excesses we remain nature ourselves. That we cannot live beyond nature because nature is all there is, and we will use up all the resources that keep us alive. But before that happens, we will have killed off the puffins and our species will spend its declining years on a planet devoid of puffins.

This is the passenger pigeon's story all over again, only this time we cannot say that we never knew, that we were unaware of the consequences of our actions. Yet the land is telling us what to do. All we have to do is to rediscover the lost art of listening to the land. Otherwise, how long do you suppose it will be before the last time you saw a puffin turns out to be the last time you saw a puffin?

◉ ◉ ◉

Craigleith slipped astern, and for the moment at least, so did the disquiet about the place of puffins in the landscape. The Bass Rock makes the instant impact of an eclipse: you can't take your eyes off it. And the passengers on this boat are not puffineers. An island of 6,000 puffins is not enough for the puffineers, and as you don't get to land on Craigleith, not close enough either. And as the prime object of the exercise of this boat trip is the Bass Rock and its top-to-toe fleece of gannets, and 150,000 gannets on a lump of rock is perhaps just too raucous and reeking, they will have gone to Farne or Lundy, Skomer or Lunga, wherever the odyssey leads them. While they still can.

The appeal of the Bass is twofold: one is the stupendously in-your-face architecture of the rock itself, its heave up from great flat breadth of the firth, its uncanny presence; the other is the mind-liberating spectacle of the gannets that assaults four of your five senses at once. (You do not want to be in a situation of having to touch a gannet. Once, about thirty years ago now, I was confronted by just such a situation; the scar took a long time to heal, of which more later.) So here, as the boat closes in, is a wedge of all-but vertical rock and sky, two thirds rock, one third sky. The rock is grey, not the red sandstone of the coastal cliffs and the castle. It is the 320 million-years-old core of a volcano, so was in its prime back when Scotland and England were colliding with each other. Think of it, the core of a volcano where nothing at all can live…becoming this. It has evolved into 75,000 gannet-sized nest sites, hacked and splintered and shattered into tiny cliffs and ledges. The rock is much more white than grey: the bright white of gannet plumage and the dull white of their droppings. The sky, a pale white, is turbulent with the flight of wide wings. These are six feet from tip to tip and pointed at the ends, the hallmark of ocean wanderers, the apotheosis of which is the albatross. It is not so surprising then, that one turned up right here on the Bass Rock in the summer of 1967, a black-browed albatross. According to the Scottish Ornithologists' Club's two-volume *The Birds of Scotland* (SOC, 2007):

> *It consorted with Northern Gannets and tolerated the close approach of humans. While the Gannets appeared to accept it and it was seen to indulge in partial display to disinterested gannets, it was frequently mobbed by gulls. It returned to the Bass Rock for about four months in 1968 and once again in 1969 for one month.*

Almost inevitably, it became known as Albert and newspapers and TV crews homed in on the Bass to enhance his celebrity. It was revived when he appeared again periodically between the early 1970s and the early 1990s at Hermaness on Unst in northmost Shetland, where he built several nests (though they remained empty) and where, strangely enough, I finally saw him. It was 1991, and I was researching a book called *Shetland: Land of the Ocean* (Baxter, 1992).

Shetland, I was discovering after visits in all four seasons, is forever dipping into a reservoir of magic tricks to play on those who go to scrutinise its landscape. At the northmost point of the northmost point of the land in April I expected big winds, maybe a flavour of something Arctic – not icebergs exactly, but something to plug me into the northness of the place. Instead, halfway across my walk out to Muckle Flugga I had stripped to the waist, and when I got to the coast opposite the lighthouse on its unforgettable rock, the ocean was almost lagoonish and gannets sat panting away the heat. I sat among puffins, burdened only by sunlight, staring in disbelief at the lighthouse, for its upended self was perfectly reflected in the water, as were hundreds of swimming puffins. I thought: "So, this is Thule?"

The gannetry was at its most aesthetically pure, no downy, scruffy chicks yet and no immatures with their weird piebald plumages, just a drifting whitewash of mature adults in pristine breeding plumage. Except... *what's that?*

It took a few moments to characterise it as a bird at all. It sat in the midst of the largest mass of birds on the rock. Any way I moved the binoculars at all, whether horizontally, vertically, diagonally or randomly, the image in the lenses simply looked like this...

gannet, gannet, gannet, gannet, gannet, gannet, gannet, gannet,
gannet, gannet, gannet, gannet, gannet, gannet, gannet, gannet,
gannet, gannet, gannet, gannet, gannet, gannet, gannet, gannet,
gannet, gannet, gannet, gannet, gannet, gannet, gannet, gannet,
gannet, gannet, gannet, gannet, gannet, gannet, gannet, gannet,
gannet, gannet, gannet, gannet. gannet, gannet, gannet, gannet,
gannet, gannet, gannet, gannet, gannet, gannet, gannet, gannet,
gannet, gannet, gannet, gannet, gannet, gannet, gannet, gannet,

gannet, gannet, gannet, gannet, gannet, gannet, gannet, gannet,
gannet, gannet, gannet, gannet, gannet, gannet, gannet, gannet,
gannet, gannet, gannet, gannet, gannet, gannet, gannet, gannet,
gannet, gannet, gannet, gannet, gannet, gannet, gannet, gannet,
gannet, gannet, gannet, gannet

...and it was repeated on and on wherever the glasses roamed across the rock, except that suddenly there was a patch of birds that looked like this...

gannet, gannet, gannet, gannet, gannet, gannet, gannet, gannet,
gannet, gannet, gannet, gannet, gannet, gannet **GIANT**
gannet, gannet, gannet, gannet, gannet, gannet, gannet, gannet,
gannet,gannet, gannet, gannet, gannet, gannet

It takes a lot of bird to make a gannet look small. In the context of a Scottish island landscape in 1991, the only thing I could think of that I had ever seen achieve something of that effect was a white-tailed eagle, but the only one of those that had been heard of in Shetland back then was almost certainly a young bird from the Norwegian gene pool that supplied all our sea eagle reintroductions and which in pursuit of some facet of racial memory we simply do not understand was trying to get back to Norway. Besides, the slumped and folded ogre at Muckle Flugga looked nothing like a sea eagle. The initial complication was while every gannet I could see was facing the same way and side-on to me, this creature was facing me head-on, and all that I could make out was a vivid yellow bill and some markings above its eyes that looked as if they were the result of some over-strenuous strokes of an eyebrow pencil. The preposterous image messing with my troubled mind perhaps could be explained by a mild case of heatstroke (I was clutching at straws again), but then the suddenly blindingly obvious illuminated the moment: a black-browed albatross. And only then did I remember all the stories and newsreels and newspaper column inches about Albert.

It stood then, towering above the heads of the sitting gannets, and as it did so, it began to unfold its wings, and these were not gannet-white but shearwater-sooty. They unfolded forever and ever. One of the most impressive aspects of gannet flight is its wingspan, all six feet of it. This wingspan was twice that length. To my intense frustration, the muckle monster of Muckle Flugga furled its wings again, sat back and resumed its patient scrutiny of the wingless creature on the mainland clifftop. That would be me, surrounded by puffins blissfully whiling away the midday sun. I suspect that's what Albert was doing, too. In the next two hours, I didn't see him move. Nor did I see a single hostile gesture towards him from the gannets. There was a vague awareness that I was witnessing a moment of some significance, at least in ornithological history.

I think the last known sighting of Albert the albatross, better known as Alby, was out on Sula Sgeir, halfway between Lewis in the Western Isles and the Faroes. But it was strange, that afternoon on a boat out of North Berwick and hard in under the cliffs of the Bass Rock, filling the glasses with that "mind-liberating spectacle of the gannets that assaults four of your five senses at once" – strange that I should think of the albatross, and my liberated mind went off on its own on voyage of memory that washed up on Muckle Flugga.

◉ ◉ ◉

As the boat neared a tall wedge of rock, the sheer number of birds washed over my mind like surf, and trying to estimate just how many occupied even this wedge of the total island was in the more hopeless realms of the fool's errand. The census is done every couple of years with an aircraft, some computers and lot of aerial photographs. Yet the restlessness of a gannetry this size and at such close quarters is positively energising. It looks chaotic at first, but the more you look, the more you see.

From a nature writer's point of view, the good thing about being on a boat that does not set passengers ashore, is that everyone on the boat is compelled to sit still and watch, and sitting still and watching is my preferred way of spending my working days. I judge the landscapes where I want to work over-time and I try to find situations within them where I sit down and grow still and watch, and try and write it down. It is very, very unusual, or at least it has been in my experience, to find another human being who is willing to share the vigil with the required degree of stillness and immersion in the land. And yet here, because the landscape has been predetermined and the only means of getting here has been taken out of the passengers' hands by the boat operators, they simply sit, face the rock, and watch. Stillness is hardly necessary: it's not like the gannets didn't see us coming. But there is a kind of agreeable camaraderie because we are all united by the common purpose of the trip. The seriousness of the photographic equipment and binoculars on display always intrigues me, and most people didn't wear binoculars and simply pointed mobile phones. Others wore back-bending rucksacks stuffed with very expensive telephoto lenses. There is – always – on bird-watching boats, an element of stuff-strutting variations on a theme of "mine's bigger than yours", but here it was minimal and relatively unintrusive. It is always hard to read how intently and analytically the other passengers are looking, but all I can do is try to memorise what strikes me, so that I can write it down in a notebook later.

In any one wedge of rock where my eye alighted, all the birds almost invariably sat facing the same way. It saves space and reduces conflicts with the neighbours. Not that there are no conflicts – there are dozens, hundreds, probably thousands of them every day, but they tend to be ritualised, a territorial defence of a sitting bird's territory of perhaps one square yard. Think about it: so many birds in such a finite and unforgiving lump of rock – there has to be a degree of organisation. Every bird must find the right square yard of rock every time it comes into land. Even when you scan the airspace above and alongside the rock, where there are flying birds in blizzard proportions, you find yourself picking out patterns.

For example: here is a piece of sky with 100 birds in it (it is a very small piece of sky). I only know this because I have a photograph of it, and I have just counted them meticulously. Firstly, all the birds except four are flying in the same direction. Secondly, all the birds except four (the same four) are flying with their wings held in a loose "W" shape. The four with a different agenda have their wings held straight and wide. All the birds nearest the edge of the rock appear to be in tight little squadrons of between six and a dozen, suggesting to me that they had all just taken off together, underpinned by collective decision. It is not far-fetched. Think how organised are their long-distance flights to and from the feeding grounds: wavetop skeins from a dozen to a hundred, regimented and efficient. Studies at the Bass Rock in around 2000–2001 found that feeding flights can be anything up to 335 miles. This capacity to extend its food searches far beyond local waters is one reason why gannets seem to be prospering while birds like puffins, fulmars and kittiwakes are struggling. There is some evidence to suggest that in addition to the essential knowledge of good fishing areas over a large area of open sea, they also follow fishing boats and work where they work. That, too, suggests a sophisticated degree of organisation.

The boat heels away from the Bass and heads back for North Berwick. The conversation level rises, variations on a theme of the day's edited highlights. All eyes now are astern. For the first time, the best view of the Bass is bathed in the best light. "Wow!" and "Look at that!" reverberate around the boat. The Bass catches fire as it subsides into the distance. The lighthouse and its adjacent long black wall (an old souvenir of garrison and prison) are as a buckle and belt across the waist of the Bass and they clasp the three essential thirds of the rock together. To the right of the lighthouse is the great vertical grey wall of the tallest cliffs, ledgeless and therefore birdless. Below the light and its wall, the bottom third is the rock's ground floor accommodation, a long and tapering wedge of wrap-round rock that reaches right to the waterline, and that is simply smothered in birds. It is a crazy paving of ledges and tiers, small fissures and buttresses. At its left-hand edge it rises past the end of the wall and connects to the Bass's most distinctive feature: its colossal dome of birds the entire width of the rock and rising to a shallow summit, a tilted snowfield of birds.

A great fan of dark clouds right across the sky is breaking up, blue fragments begin to burst through. The airborne hordes of gannets are the first to catch the sun, then the light spreads across the crown then down over the bright white of the lighthouse buildings, then the ground floor begins to glow. The sun chases down the boat's wake and finally that ribbon of dazzle is all that binds us to the receding gannet realm that can so liberate a human mind.

◉ ◉ ◉

I mentioned having once been confronted by a situation where physical contact with a gannet appeared to be inevitable. It was a Bass Rock gannet. The situation arose about thirty years ago when I went for a walk out from Aberlady Bay along the beach towards Gullane, just a few miles to the west of North Berwick. I came over the dunes that lie on the landward side of the sandy beach and my eye alighted on something white at the edge of the tide. It moved awkwardly. In the binoculars it resolved into a gannet. Gannets don't walk along sandy beaches. They have no reason to. Other than an occasional step to resolve a nesting season spat on the Bass, I can't imagine them walking anywhere at all. They deal in the realms of water and air, and they have no reason for land other than to lay an egg on it every year. So it is an unaccustomed and unwilling walker at the best of times. This turned out to be not one of the best of times.

It was a weekday and the beach was empty. I was on my own. One of the gannet's wings was half open and something pale green was caught there. I walked slowly in a wide semi-circle to try and make my approach as unthreatening as possible. I was thinking it might somehow avoid stressing the bird but it would prove rather too late for that. The pale green thing streaming out from the apparently damaged wing was a piece of fishing net. As soon as I realised that my heart sank. The piece of net was not just caught around its left wing, its left foot was ensnared too, with the result that every time it moved its foot the net tightened and bit deeper into the leading edge of the wing. I could see now where the blood seeped.

I sat down on the sand and had a good look through the glasses. It seemed just possible that if I could get hold of the trailing piece of net, I might be able to free the bird's foot. Then, it also seemed just possible that the bird itself might be able to manoeuvre the net from the cut in its wing and out towards the narrowing wingtips, and finally be free of it. Whether or not it would then be able to fly was a completely different question. The chances of the gannet sitting still and docile while I tried to get the net off were not high. Now, a quick call on a mobile phone would have brought a swift response from an SSPCA worker in a boat or a 4x4. Thirty years ago, I didn't walk with a phone in my pocket.

Get this over with. A bold approach produced a predictable flapping and painful retreat from the gannet, but I persisted, talking in a hopefully reassuring voice, until I got hold of the net. For a few moments I walked behind and slightly to one side of the gannet as it tried to get back to the water. It occurred to me that from a distance it would have looked like I was taking a gannet for a walk on a kind of leash. Its struggles had succeeded only in looping the net round its huge webbed foot twice. Improbably, by flicking the loose end of the net as it moved its foot, I freed up one of the loops but the other was tighter.

The bird stopped. I crouched by its tail and eased a hand towards the still-trapped foot. I was within an inch or two of getting hold of the square of mesh that held its foot when the bird turned its head towards me and that beak that crushes the life out of fish for a living grabbed my bare wrist and let go again a couple of seconds later. The pain was extraordinary.

Three things happened at once.

One – I let go of the net involuntarily.

Two – the gannet made for the water still towing the net, uselessly flapping its wings, and contriving quite by accident to free its own foot as it fled, but not the net from its wing.

Three – a neat bracelet fashioned entirely from beads of my own blood appeared magically on my left wrist.

I watched the bird as it beat the water with its wings and travelled about fifty yards offshore. I reasoned that at least it was safer there than on the beach, where sooner or later it would lose a fight with someone's pet dog. If it did manage to extricate the net from the cut in the wing, then extricate the entire wing from the net, perhaps it would heal out on the water.

When I examined my own wound I saw that the bird had punctured the skin in twelve places round my wrist. They were all shallow, but they all bled and the cumulative effect was really quite impressive. The following day I was in a shop and when I handed over the money, I did so with my left hand, and too late, realised that the assistant was staring in some horror at what looked like an unsuccessful attempt at slashing my own wrist. I thought about trying to explain, something beginning: "I was bitten by a gannet when I tried to help it along the beach…" and then I thought better of the idea. I wore the bracelet for weeks before it faded. The memory is no less vivid today, thirty years after the event. It never faded, likewise the memory of the pain.

Part Three

Smoke Signals

Chapter Ten

Touchstones

INTERNATIONAL AIR TRAVEL is not my thing. Quite apart from the ideological and ecological considerations – these are as huge and black as thunderclouds, as grim and darkly portentous as a skein of Airbus 380s putrefying the air over Heathrow awaiting permission to land (the Airbus 380's website croons "the largest commercial aircraft flying today" as if that was a good thing, before adding that its wingspan is 260 feet and nine inches) – quite apart from these considerations, big airports intimidate the shit out of me, not least because they constitute arguably the most unnatural habitat on Earth. When it comes to air travel, as a general rule I prefer wee planes and wee airports. Wee airports like Foula in westmost Shetland, where, the last time I looked, the terminal building was a shed full of broken television sets; wee airports like Papa Westray, where the shortest scheduled flight in the world puts down 1.7 miles from Westray (total time, including taxiing, around two minutes, give or take the odd fluctuation for wind direction); wee airports like Fair Isle, where I once met the woman who drove the airport fire engine (and ran the post office) and whose duties included driving the length of the runway with siren blaring to clear it of sheep just before a plane was due.

Float plane docks are best. I like wee planes that leave a wake when they land or take off. I acknowledge that I could not have got to the float plane dock on Kodiak in Alaska's Aleutian Islands in the summer of 1998 without succumbing to the multiple indignities of being treated like a rugby ball in an All Blacks scrum through four large airports: Edinburgh, London Heathrow, Minneapolis and Seattle, before the comparatively modest Juneau put me down gently after twenty-four hours of hell on Earth, and off it for that matter. But once in Alaska (on a BBC radio ploy to make two programmes about the relationship between people and wilderness), I was treated over the course of three weeks to a masterclass in the amiable art of the bush pilot, the acceptable face of air travel. Thus, I was put in positions from which I would eyeball a grizzly bear at twenty paces, a humpback whale at not much more than touching distance, and I would redefine the possibilities of the very word "wild". It may be more than twenty years ago now but I am still refining my response to what I saw and felt there.

The wholly negative byproduct was that it put me off long-haul flight forever, this despite the fact that the experience of actually being there changed my life, made me re-evaluate my work as a nature writer. I have reasoned subsequently that even if I only ever work within my native Scotland, with occasional forays into wilder tracts of the north of England, my horizons as a writer would still be immensely broadened because of that exposure to pure wilderness and the kind of creatures that made it function. It has also added an indefinable edge to the instinctive north-ness to which my writing has always inclined. So Alaska became a touchstone for me, and the fact that echoes of what I found there reached me on two other BBC projects in Iceland and Norway, simply reinforced that instinct.

Touchstone: it's a lovely word. It works at many different levels. It can mean something the size of Alaska, or it can mean a pebble that fits the palm of my hand in circumstances that provide connections. I like to pick up stones, especially from the bed of a burn or a shoreline. I like to rub my fingers around them, wonder where they began life, feel the heft of them and the zest of them in my hand. Usually I put them back where I found them, but not always. Right now, there are four stones that have been with me for varying periods from decades to a single year, three in the room where I work and one in the car. One came from Iona, one from within a few yards of Gavin Maxwell's home at Sandaig, the West Highland shoreline he called Camusfearna and where he wrote *Ring of Bright Water* (a book that electrified me when I first read it aged somewhere between eighteen and twenty and made me think I would like to write like that), one that came from the beach by St Ninian's Cave on the Galloway coast of south-west Scotland, and – the newcomer of the group – one from a small glen in the Balquhidder hills of the Loch Lomond and the Trossachs National Park.

Their stories are these:

The Iona stone is dark grey, almost black, and looks a bit like a standing stone except that all its edges are softly curved. It also has a rounded bulge to what I think of as its lower half, this perspective being achieved by the way it fits most snugly into my hand. It looks pregnant, as if it might give birth at any moment to a little rounded boulder, which – in time, in time – might grow into a standing stone itself. It is unexceptional to look at, but it fits into my hand so agreeably that it decided to stay, and I decided to let it. In all my island travels covering the better part of fifty years, Iona imparts a unique essence. I could not begin to define what it is. I only know it is so.

"A test or criterion for determining the genuineness of something," is how my dictionary defines a touchstone. It seems particularly relevant to the stone from Iona. The Sandaig stone is the smallest, just over two inches across, an inch high and an inch thick. It lay near the stone that commemorates Gavin Maxwell's otter Edal. There are words on the "front" of it, words I copied out from Maxwell's dedication in his book *Raven Seek Thy Brother*: "EDAL: Whatever joy you had from her, give back to nature." It is always cold whenever I pick it up, regardless of the temperature in the room. I occasionally go over the letters with a black pen, because rolling it around in my hand blurs them from time to time, and I admire and approve of the sentiment so I keep it legible. It puts me back on that West Highland shore in a second.

The St Ninian's Cave stone almost shouted my name out loud at me. A characteristic of the stones on that beach is that they are dark grey but weathered with white shapes into abstract art of infinite variety – all of them abstract that is, except one. The one in question, the one that summoned me, has an almost perfect and appropriately white representation of a walking polar bear, and that is why it demanded to come home with me. It was only after I had admired it for several minutes that I realised there was more to it. On the other side is a breaching whale complete with a cloud of "breath" coming from the top of its head. Whales and polar bears. A fragment of Alaska washed up on an eastern shore of the Atlantic Ocean.

The fourth stone is the one from the Balquhidder hills, and it travels in the car at the moment. They have all taken turns at travelling with me as a kind of St Christopher's talisman. This one came from the stones on the inside of a bend of the burn that flows through a glen beneath a golden eagle eyrie, and which has been the centrepiece of my work with golden eagles for more than forty years. Two things attracted me to the stone. One was simply that it glittered. The other was more startling, for in outline it is an almost perfect eagle head. It even has a hooded "eye". And when I found it, it was lying less than half a mile from the eyrie buttress.

These, then, are the touchstones of old summers that have grown into touchstones for the nature of all my summers. The definition of touchstone quoted above is actually the second one listed in my increasingly decrepit dictionary. The first one is this: "Black flint-like siliceous stone that when rubbed by gold or silver shows a streak of colour… formerly used to test the purity of those metals." If you had asked me before I looked up the word "touchstone" what colour my eagle stone glitters, I would have said like a mixture of gold and silver, depending on the light.

◉ ◉ ◉

The symbolism grows and deepens with the years. Other stones in other landscapes drift through my hands. Some I have kept for a while and then put them back, in the original landscape or a new one. These four have stayed longer than most, I suppose because in my mind and the circumstances of my finding them, each perfectly fulfils the idea of determining the genuineness of something.

The eagle stone from the eagle glen has added to its status when a white-tailed eagle – a sea eagle – began to turn up in the same glen, a wandering youngster doubtless attracted by the presence of golden eagles. When I wrote an earlier book called *The Eagle's Way* (Saraband, 2014) I had advanced then developed a theory that followed the twenty-first-century reintroduction of sea eagles into Scotland's east coast at the Tay estuary. It was that a certain innate instinct of race memory – or something like it – had propelled some of the young birds sourced from Norway's island-strewn west coast to head straight across country from the Tay estuary to Mull, where there has been a thriving population of sea eagles more or less since west coast reintroductions began in 1975. Why did it happen? My best guess was that Norway doesn't have an east coast, that that instinct was ingrained enough in some individuals (and eagles are among the most individualistic of all birds) to persuade them to seek out an island-strewn coast that faces the sunset, as opposed to the east coast whence the sun *rises*, and which is more or less devoid of islands. In their travels (and all young eagles are great travellers) they met young wandering golden eagles, and began to keep each other company. Thus, "the eagle's way" between east coast and west became a kind of two-way highway for both species of eagles. I posed the obvious question: new behaviour, or very old behaviour indeed and the circumstances that permitted it to recur were cemented into place only by the east coast reintroduction of sea eagles? From what I have learned since, including the tendency of some recolonising sea eagles to choose historical nest sites not occupied in more than a century, I incline towards the latter.

So, when I say that Alaska became my touchstone, and that it also reinforced the instinctive sense of northness that attends my work, and that I sensed echoes of it from time spent in Iceland and Norway, I should perhaps have added that I had assumed ever since Alaska that nothing could match its impact on my nature writer's sensibilities. What could possibly surpass it other than to go back and travel there under my own steam, following my own nose rather than following a pre-determined BBC schedule? That question lay on a stoorie old shelf at the back of my mind for years, and I never troubled to blow the dust off it because of the sheer cost and hassle of getting there, not to mention my ideological scruples about long-haul air travel. Not only could I not afford it, more important by far is that the planet cannot afford it.

But then a warm wind of opportunity blew my way, not to go back to Alaska, but to Norway, and not to mainland Norway but to somewhere that would surely slake my twin thirsts for islandness and northness: Lofoten.

Lofoten is Norway's Outer Hebrides, a chain of islands off the north-west coast but with two distinct differences from the Outer Hebrides. One is that every island looks like the Skye Cuillin, its airspace so crammed with mountains that at first glance you wonder where they put the towns. The other is that they lie inside the Arctic Circle. I had already decided that the nature of summer in Lofoten would be different from all the summers I had ever known. I would turn out to be right, but I was quite unprepared for the nature of that difference, and I had misinterpreted what the word "Arctic" was to entail in the second decade of the twenty-first century.

◉ ◉ ◉

My interest in Lofoten, apart from being an enigmatic chain of islands with a distinctly northern cast, had been quickened by that slow and sensational spread of sea eagles across much of the map of Scotland. Lofoten is at the heart of an area of northern Norway with a healthy population of sea eagles and the source of those young pioneering eaglets that were the key to Scotland's original reintroduction project in 1975. Touchstones, you see?

From where I live, getting to Norway is easy. At the right time of the day you can do Edinburgh airport from here in forty-five minutes. Bergen is a two-hour flight. But because of a string of logistics that need not detain us here, I had to get a train to London, a tube to Heathrow, an overnight stay in a hotel there, and only then, after having travelled 400 miles *south*, the journey north to the south of Norway could begin. Then there was the internal flight to the point where a hire car would complete the journey into the islands. The words "tube" and "Heathrow" still have a power to chill. For entities that were specifically designed to transport human beings (a) around London and (b) around the rest of the world, their twenty-first-century manifestations contrive to be inhuman in every aspect. This, I told myself as I tholed their heaped indignities, had better be good.

Chapter Eleven

The Land of Havørn (I) – Under the Blue Mountain

THE WORD IS *havørn.* Literally, "sea eagle", when the sea in question is Norwegian and it quivers with islands tumultuous with mountains. And now I travel back there in my mind, oh...so often. And in my mind, in my mind's eye, and for that matter in my mind's heart, there constantly reappears a particular mountain. It is just one of hundreds by which Lofoten stamps its identity, its temperament of storms, its uncategorisable beauty, upon the consciousness of the susceptible travelling mind. Mine was one of the susceptible ones. That singular mountain made something of a habit of rising before me again and again throughout my time there. It seems to me now that whatever my direction of travel, there it was, casting its shadow, barring my way or staring down at me as I sidestepped it deferentially, imposing its aura, impressing itself upon on me.

What was special about it? It wasn't tall or pointed or razor-edged as so many of Lofoten's summits are. But if you have an imagination like mine, it takes the shape of an eagle, a huge eagle head thrust forward from a pair of mantled wings (these are mountain ridges that rise slightly at the "elbow" then bend almost at right angles, as mantled wings do, then droop almost vertically in folds and gullies that suggest illimitable depths of feathers such as are revealed if you ever see the chest and belly of a perched sea eagle tormented by big winds). The chances are you go to Norway with preconceptions about the natives' taste for mythology (and come back with them, too: trolls glower at you from every tourist shop window) and a particularly vigorous tribe of Norse gods. If your imagination ever strays – as mine occasionally does – back to that long-lost era of worship that placed its gods on mountaintops all across the northern hemisphere, then surely here was an apotheosis among mountain gods. So here was a mountain that was surely primed to haunt a susceptible mortal like me. So the mountain and the bird that in a roundabout way had lured me here have become emblematic of that astounding landscape, in the same way that for Alaska twenty summers before it was a grizzly bear and a humpback whale.

The connecting flight north up the Norwegian mainland from Oslo to Narvik was the part of the journey that moved from one planet to another. Looking down on glacier-strewn summer snowfields and mountain summits from 30,000 feet, two thoughts occurred to me. One was that somewhere down there, the invisible line we call the Arctic Circle had just been consigned to my south, and that had never happened before. By the time the aeroplane reached Narvik, it would be 120 miles to my south.

The other thought drifted back to the whole question of flight – of the vexing human addiction to air travel despite everything we now know about its consequences for the planet, and, because of the nature of the enterprise at hand, sea eagle flight. Without having to climb to 30,000 feet or anything like it, without having to seek permission to take off and land, without consuming so much as a single millilitre of aviation fuel, and in complete silence, a sea eagle could make the same journey this plane was making, enjoy the same view of the mountains and their glaciers, except that it would be unconstrained by the frame of a cabin window. It would take much longer, of course, but as a result of its flight the air through which it passed would be utterly unimpaired. A sea eagle wingspan is large, at least by bird standards: a big female can nudge up towards eight feet, or 252 feet and nine inches less than an Airbus, not that I was flying in an Airbus to Narvik, but you get the gist. Flight as designed by nature is an exquisite phenomenon. Flight as reinvented by humankind is more or less grotesque. Such was the conflict that troubled me.

It's not that I do this very often: my carbon footprint is a light one. But somehow it all felt so much more intrusive in somewhere like Arctic Norway. It is impossible, of course, not to be seduced by the view of glaciers and summer snowfields from 30,000 feet, but just in case there was the slightest chance of slipping into a complacent frame of mind that assumed all was well with the Arctic world, the captain of the aeroplane started talking to us as he prepared to land in Narvik. He spoke first in Norwegian, but it was obvious at once even to non-Norwegian speakers that he had just said something remarkable, such as the immediate and animated response all through the cabin. Then he spoke in English, concluding the standard landing script with the non-standard bit: "...and the temperature on the ground is…ooh, thirty-four degrees Celsius".

"Did he say twenty-four degrees?"

"No, I think he said thirty-four degrees."

He *did* say thirty-four degrees. I did some quick mental arithmetic: ninety-two degrees in old money, 120 miles inside the Arctic Circle. Oh yes, the system is well and truly broken. On the tarmac, the heat was like a wall. In the airport, the man from the car hire firm said:

"Hi, welcome to Spain."

That week, Norway and Sweden would record their hottest ever temperatures. A climatologist talked on the radio about "a dome of heat stationary over northern Europe". The nature of summer in the far north of the world had just lurched into uncharted territory.

The drive from the airport deep into the islands was bridge after bridge, causeway upon causeway, mountains that leapt from seas and fjords and lakes and other mountains. When the car finally stopped I wasn't so much travel-sick as sick of travelling, but then I discovered that I had come to rest in a magic land, a tiny village that clung to a mountain-rimmed fjord.

I was having trouble coping with the day's hoard of sensations, and just when I thought there could be nothing left but troubled sleep, I remembered our small party hadn't eaten for quite a long time, and we ate smoked salmon and French bread and drank wine on a tiny terrace outside a timber cabin. It was 11-o'clock-going-on-midnight and the sun shone. It was somewhere about then that I learned the working definition of the phrase "Arctic Circle". It has nothing whatever to do with land or cold, and everything to do with light. It is a mathematical line drawn at latitude sixty-six degrees and thirty minutes North. Its purpose is to mark on a map of the north of the world the southern limit of that zone wherein there is at least one period of twenty-four hours every year during which the sun does not set, and another during which it does not rise, and it is as simple as that. That knowledge made things a little easier to understand as the evening stretched on and on into the wee small hours and the light scarcely dimmed. The day finally came to an end when I simply let my heart and mind and travel-weary body find what felt instantly like the most natural of homecomings, wreathed in beauty of a particularly rare order, and in unaccustomed warmth.

◉ ◉ ◉

The first morning after the night before began at a tiny nowhere on the map called Eggum, which, mysteriously, seemed to infiltrate my mind with a Yorkshire accent. It is an end-of-the-road place on the seaward edge of the island of Vestvågøy, and beyond the end of the road there was a track that threaded a narrow shelf of land between a wall of mountains and the Norwegian Sea. Might that, I wondered, be a fair description of almost every other road in Lofoten?

We parked under a mountain at the end of the road. Always, it seemed, as that week unfolded, we parked under a mountain. Always with mountains like these, not high but as sharply defined as the Skye Cuillin, they demand that you look up first. Clouds had crept in during what passes for night at these rarefied latitudes of summer, and as yet they still clung to summits, although such a sun and such a heat was about to burst them apart, annihilate them. Where the ridge disappeared into the cloud for the moment, there was the expedition's first sea eagle. I hadn't even had time to put my boots on. A small cluster of ravens clamoured around it, voices softened by distance, adrift among the thermals, falsetto, throaty, harsh and sweet, deadly serious and comic.

Did they think they might embarrass a sea eagle? Or did they even think? Is this just ritual, the clamour for the sake of the clamour, with all the effectiveness of farting against thunder? She (the sea eagle, hugely female) contrived a back flip to present talons upside down at the posse of irritants. I don't know exactly how that must have looked from a yard away, nor how unsettling it must have been for raven flight given the air currents that must have flowed from such wings as they executed such a manoeuvre. Almost at once she righted herself again by completing a barrel roll and suddenly she was alone in the sky and in possession of precisely the same piece of mountainside as she had been when the ravens turned up. It was as if she had simply assessed the opposition, called up the one gesture she calculated would be necessary to deal with it and executed the gesture while apparently resting on a cushion of warm air. The ravens' instant, scattered departure suggested they knew they were outclassed, but there again, they almost certainly knew that before they moved in. Perhaps it is simply that in the evolution of the relationship between sea eagle and raven, they have come to the conclusion that to irritate is better than doing nothing at all, that they must fulfil a stipulated principle of the natural order. Ravens are spectacular fliers in their own right, but inevitably, they lack the grand gesture and the raw power of an eagle.

By any standards, the sea eagle's manoeuvre was very impressive – any standards that is except one. I have watched my fair share of meetings between sea eagles and golden eagles in Scotland's island west, and while it is true that a sea eagle will usually out-muscle a golden eagle on the ground and drive it away from a kill, if the confrontation is in the air I have yet to see the sea eagle that can match the golden eagle's apparently limitless repertoire of the art of flight. They are out-flown and out-thought every time. In any other company other than golden eagles, the sea eagle excels against all-comers, and it can certainly afford to be contemptuous of ravens.

I tried out a couple of mountain names in my head: the mountain where the sea eagle held sway was called Mustaren, an outcrop of a ridge at 400 metres on the way up to Jellvollstinden at 746 metres. This gave me an immediate yardstick to work with, for the mountains, all of which rose from the sea, *looked* much higher. The midwinter imagery beloved of Norwegian tourist brochures and climbing guides reassert their essential Arctic-ness of which this preposterous heat-dome had stripped them. In that midwinter guise they looked as imposing and shapely as the Swiss Alps. With new landscapes, especially landscapes rooted in seas and flooded with twenty-four hours of daylight, you have to make adjustments of scale, and it takes a while.

Something else fell into place as I watched the sea eagle cruise the ridge. Throughout the length and breadth of the Lofoten archipelago there must be hundreds of such mountain ridges, their cliffs wide open to seas and fjords and lakes, hundreds of suitable nest sites for sea eagles, hundreds of places to fish and catch seabirds, and as a result, surely hundreds of sea eagles. Norway harbours more than 2,000 pairs, more than ten times as many as Scotland. I hazarded a hasty guess, which more considered deliberation has reinforced, that Lofoten alone surely exceeds the Scottish population. So what fell into place that first morning under a modest mountain called Mustaren was a clear understanding of why a sea eagle raised on Scotland's east coast might seek to reclaim its west coast Norwegian ancestry by crossing the breadth of mainland Scotland: in pursuit of some instinctive quest for a land of mountainous islands that eagle would naturally end up on Mull, the Small Isles, Skye (perhaps especially Skye with the Lofoten-esque Cuillin mountains at its heart).

A second essential truth I had previously only guessed at was now confirmed for me, too. In Lofoten, and for that matter throughout Norway, sea eagles comfortably outnumber golden eagles, a situation that was the historical norm in Scotland until the Victorians embarked on a campaign to blast the sea eagle from the face of the land, a campaign that was all too successful. But the growth of the sea eagle population in Scotland is now such that sometime in the next twenty years the historical norm will reassert itself and the sea eagle will once again outnumber the golden eagle in Scotland, too. The situation should trouble no one, for the two species thrived side by side for many millennia, as they do today in Norway, and they resolve their differences effectively; and as we have seen and are seeing increasingly, the young of both species are happy to share the same airspace as they criss-cross the land, all its coasts and all its heartlands.

Having relished the company of eagles of both species for many years now and watched the sea eagle reintroduction project with eager fascination throughout all its phases, it was the sea eagle tribe that put Lofoten on my personal map of the world in the first place, and here was the best of all omens for my time in Lofoten in the first hour of the first morning on the first day. The eagle veered below the ridge and was lost to me, but I sensed that I had just found a new relationship in nature that was utterly true and life-enhancing, and that would never leave me. I hugged the knowledge close, for its quality was as rare as it was precious to me. I turned and headed south down the coast towards Utdalen, south down the outermost edge of outermost Arctic Norway.

A lake lay in the lap of two curving mountain ridges, each of which climbed to cliff-girt, cloud-nudging summits before swooping a thousand feet to collide beautifully at a low mountain pass that I would call a *beallach* at home. It is the Gaels' word. The landscape of north-west Scotland, and especially the Hebrides, is named by two languages, Gaelic and Norse, for the Vikings made their presence felt there, too, and left their fingerprints all over the map of the land. Again and again, the landscapes of Lofoten would stop me in my tracks with echoes of the Hebrides.

The heatwave held, slowing the pace; the morning was still, the lake lay as flat as a sheet of newly forged steel, dark greyish blue that evolved into dark bluish grey, and only paling where a band across the middle of the water caught a hint of a matching band of white cloud high above the summits. The grassy shore was thronged with tall buttercups and orchids. All that moved at that moment were slow clouds, and on the water, a slower and apparently solitary black-throated diver.

So that first walking mile inside the Arctic Circle was lathered in sun cream, graced by eagle and diver and wildflowers and what felt at first glance like a gasp-out-loud, out-of-this-world landscape, except that in this particular corner of the northern world it would be replicated again and again and again. The combination of lakes, fjords and open sea strewn with apparently endless variations on a theme of near and far mountainous islands and lined to the east with the higher, bulkier mountains of mainland Norway…that would recur relentlessly and become the signature-in-landscape of the whole adventure. That, and for the moment at least, a climate that appeared to have migrated from the Med. The black-throated diver was so indolent that it was tempting to believe it was sunbathing. Such conditions were as unfamiliar to Arctic-born birds as they were to Arctic-born humans, and neither species appeared to have had any more idea about what to do with them than the other.

I tried running the name of the lake around in my head, Nedre Heimredalsvatnet, and the *vatnet* suffix occurred often enough for me to conclude that it means "lake". The particularly striking mountain that rose from the south-east corner of the lake and soared from the *beallach* was the north face of Blåtinden. We were destined to meet again and in equally alluring circumstances later in the day. For the moment, I was close enough for a black-throated diver to fill my binoculars, the bird idling and utterly indifferent to my blatant presence on the shore. From sooty-black stem to bright white stern, the most exquisite bird plumage in the north of the world blazed in sunlight. It was the second time in the course of writing this quartet of books that I had been in the right place at the right time to witness this dream of a bird at close quarters, the first time being on Loch Tulla in the Black Mount hills of Argyll. I recalled then the wildlife writer Mike Tomkies's assessment of the bird in his book *A Last Wild Place* (Cape, 1984) and it bears retelling here:

> *For me they embody as no other creature does the wild spirit of the loch…when flying high on their short but powerful wings they look like arrows of twanging steel…Their magnificent summer plumage, with its rich blend of slate greys, blue greys, purples, blacks, creamy underparts and sooty throat patches must be the sleekest of all water birds', and their array of intricate white neck stripes and snowy wing bars, if you're lucky enough to get close, dazzle the eye.*

"The wild spirit of the loch" on the waters of Nedre Heimredalsvatnet turned towards me and drifted closer to the shore, and sure enough, the closer it came, the more the eye was dazzled, the more unfettered my admiration of the creature that, for Mike, was a kind of emblem of wildness on the long West Highland loch that washed ashore a few yards below the cottage where he did his best work. He had renamed the place Wildernesse.

The founding father of all contemporary Scottish nature

writing, Seton Gordon, was deeper into the Norwegian Arctic when he recorded a historic encounter with a black- throated diver on Prins Karls Forland (which he rendered as the very un-Norwegian-sounding Prince Charles Foreland), Spitzbergen, in July 1921:

> *Flying at great speed, and slanting towards the water, there came a dark, diver-like bird which settled on the lagoon near us. Through a powerful glass, and after careful watching, it was identified as a black-throated diver…the first definite recording for Spitzbergen. It probably nested later by one of the hill tarns, as yet ice-bound, in the interior of the Foreland.*

He was writing in his book *Amid Snowy Wastes* all but 100 years ago, and decades before the concept of global warming had reared its troublesome spectre. He described an utterly different Arctic world from the one where I walked. The tarns were ice-bound, it was still snowing, and "the snow on the Foreland was still almost continuous…Winter often descends on Spitzbergen before September arrives, so Prince Charles Foreland must be under almost eternal snow." Ah, but that was then. Now, whatever happened to the concept of eternal snow, even in Arctic Norway?

It occurred to me suddenly that he was making his first trip to the Arctic, as the photographer on an Oxford University research trip to Spitzbergen. He was a student, passionately interested in the wildlife and landscape of the nearest thing Scotland has to the Arctic, the plateau lands of the high Cairngorms. And the black-throated diver would have been as a bond to the other landscape love of his life, the Hebrides. Almost a hundred years later, I was making my own first trip to the Arctic, and in our different ways we had both seen and responded keenly to a single black-throated diver. I never tire of Seton Gordon's books, and I am always delighted to feel some kind of connection, however vigorous or tenuous, to his trail-blazing nature writing. For a moment I was amazed that it should happen north of the Arctic Circle. Then I thought:

"But of course it would happen here. Why wouldn't it?"

It was an extravagant setting to frame a single black-throated diver: the tranquil loch, the two mountains beyond its far shore, each dipping a mighty shoulder towards the other, a fraternal clasp, the warmth, the tranquillity, the verdant green that smothered the lower half of the mountains and reached high up into gullies and surrounded buttresses. This was not the Arctic I had believed in. Seton Gordon's iced tarns and eternal snow were what I had expected, even though I am well read about the crisis in Arctic ice and Greenland's disappearing glaciers, with the extent of the melting transmitted in billions, even trillions of tons of water. It took this moment of bearing witness to drive the message home. And do you know what was the most sinister aspect of the evidence confronting my own eyes? It was so utterly beautiful. It had never occurred to me that climate's cataclysm was so two-faced.

In my head I heard again the voices that had introduced me to this new reality, voices both incredulous and uncomprehending...the two passengers as the plane prepared to land:

"Did he say twenty-four degrees?"

"No, I think he said thirty-four degrees."

And the car hire man:

"Hi, welcome to Spain."

The heat on the tarmac that was like a wall.

If it was any cooler out on these islands, it was a matter of a couple of degrees. And a black-throated diver looked like it was wearing too many clothes. Behind my back, the Norwegian Sea sprawled, flat and silent and blue, every shade of blue you could think of, and every other island and all their mountains paled into the distance from almost navy blue for the closest ones to a kind of bluish off-white for the most distant. It was an extraordinary vision, and it was the most disquieting mask I have ever seen.

A pair of red-throated divers with two young were calling on a small fjord. In Scotland the red-throats nest in small lochans near the coast and fly down to the sea to feed. Black-throats nest on bigger lochs inland and fish where they nest. But here in Lofoten there is no "inland". The islands are long and narrow and stuffed with mountains and ragged-edged with fiords. In my mind's eye, red-throated divers are Shetland, Orkney, south Skye, Raasay; black-throated divers are the Trossachs, Highland Perthshire, Loch Tulla in the Black Mount mountains at the edge of Rannoch Moor, Loch Shiel...and it is only in winter when the twain shall meet in the firths of the east coast like Forth and Tay, and in drab winter plumage at that.

For me, and thanks mainly to Shetland and one unforgettable night sleeping out on Raasay, it is the red-throated diver that is the embodiment of a particular strain of wildness, and while it does not "dazzle the eye" (to borrow the phrase from Mike Tomkies) there is a sleek silkiness to the red-throat, and the fact that the colour of that exquisite throat patch is matched in its neck stripes and in its extraordinary red eyes, lends it an almost unearthly distinction. And when the down-curving, far-carrying calls of the two adult birds synchronise in unbroken rhythmic call-and-response for minutes on end, it sounds like the very breath of the wildest of landscapes, like nature breathing. My first Lofoten day was piling one wilderness sensation upon another. It was dizzying, bewildering, overwhelming, and unrelentingly beautiful. And still the day was far from done with me. I was about to be reacquainted with Blåtinden, whose north face accommodated the black-throat's lake.

Its south-east face not only embraced a wide corrie with a high, hidden lochan, but it also towered over a roadside lake called Keilvatnet. For more than half my life now I have greeted the autumn arrival in Scotland of wild whooper swans from Iceland and Scandinavia as a presence that hones the most cherished wildnesses of my country with a keen Arctic edge. And of all those manifestations of wildness, of all those tribes of nature that define all nature's seasons, there is nothing – no wild presence of any kind – that has got so deeply under my skin and defined more truly my relationship with nature for fifty years. To my mind, whooper swans embody like nothing else my sense of wildness, of flight, and of northness. And there, on a roadside lake called Keilvatnet under a mountain called Blåtinden, which also gives succour to black-throated diver and sea eagle, there were two adult whooper swans and a very young brood of six cygnets, and that was a thing I had never seen before, and my heart turned over.

I had been to whooper swan nesting grounds before – in Iceland, where I watched them on the black-sand shores of a lake with the mighty volcano of Hecla for a backdrop. I was making a television programme about swans for the BBC, and one sequence appeared to show me getting quietly emotional at what I called "the missing piece falling into place", watching newly hatched whooper swan cygnets appear from beneath the feathers of the pen for the first time. The truth is that only the cameraman was permitted to stay on for as long as it took for the eggs to hatch, and my part in the sequence was faked. By the time they hatched I was about a thousand miles away, for I had been back in Scotland for several days. The experience put me off wildlife documentary television for life.

Later in the film, there is a sequence in which the fully fledged cygnets take their first flight. But it was filmed in Scotland using hand-reared birds. I was asked if there was "anywhere near Stirling that could pass for Iceland". I took the crew to a small, peaty lochan on a high moor just a few miles from where I live. The subsequent sequence was beautiful (it was filmed by Simon King, no less) and would have deceived every viewer. But the deception rankled with me, as deceptions always do, and it still does; witness the fact that here I am reliving the event twenty-five years later. I had never lost sleep about it, and at the time I was hopelessly naïve about television film-making. But from time to time, usually while watching a small group of north-making whooper swans on a small lochan of south Skye or north Mull some April or other, knowing they were preparing for the long haul back up the north Atlantic to nesting grounds in Iceland or Norway, I have thought how much I would like to lay those irritating demons to rest, to see wild cygnets on their native waters, and to be in a position to write it down. And suddenly, while my mind was on something else altogether, there they were, one late afternoon of midsummer on an island off the north-west coast of Norway. They looked to be about three weeks old, and they would have a lot of growing to do before late September or early October, when they would follow their parents back down the Atlantic or the North Sea, but twenty-four hours of daylight assists the process, with the possibility for twenty-four hours of feeding. Nature rarely leaves such things to chance.

The swans were at ease. The adults would never have known such weather before. The cygnets were born into an extraordinary season. But that demure image the family presented to the world was another deception. The very coupling of the words "swan lake" is a deception. For these were as much birds of that mountain as they were birds of that lake. The adults would know the mountain as well as they knew the lake, how to fly over it and navigate through its upper slopes, how to find a way through the contours and spaces between its cliffs and buttresses, the upthrusts and gullies of its ridges; they would know its high lochan hidden in the embrace of its south-east facing corrie, the walls still flecked with the last of the snow. Swans like these have no fear of mountains, no reason to shun them, for they are among the strongest of fliers and their migration routes vary from wavetop level to the edge of the jet stream. Navigating mountain ranges and archipelagos is second nature to them, arguably first nature.

Often, when I encounter a family group of whooper swans newly arrived in Scotland from Iceland or Norway, I think about the younger birds, in particular – no more than three or four months old, their first 1,000-mile migration behind them already, still grey and improbably slim compared to the adults, still lacking the yellow blaze of the adult bill and the mellow muted brass voice – and I wonder what the landscape is like where they have come from. Now I can answer my own questions, now the final piece *is* in place.

Throughout the autumn and winter and early spring that followed that extraordinary Lofoten summer, I scanned all the whooper swans I saw: I seek them out on the reliable watersheets where they turn up year after year. These would include Loch Leven in Kinross, a national nature reserve and a gathering place for wildfowl and waders, and where whooper swan numbers often build up to over 200 before they scatter more widely across the land; Balquhidder Glen in the Loch Lomond and the Trossachs National Park, where the watersheets are hemmed in by steep mountainsides; its near neighbour Loch Lubnaig, where for more than thirty years now I have watched truly wild mute swans dish out a harsh welcome to whooper swans that pause there; Loch Dochart, a watershed further north from the Balquhidder lochs and with a harder-edged Highland feel; the Lowland fields of the Carse of Stirling just to the south of the first and last mountains of the Highland Edge.

Then there is Caerlaverock on the Solway coast, Scotland's most southerly shore, where they gather at a Wildfowl & Wetlands Trust reserve and are fed daily by reserve staff; the whoopers join the expectant throng of mute swans and geese and ducks in response to carefully timed calls and whistles from the staff before food is thrown to them from a wheelbarrow while human visitors come to watch the spectacle from a purpose-built hide. I don't much care for that part, for it makes wild Arctic swans behave abnormally and turns them into performers to gratify something that strikes me as less than admirable (the trust does all kinds of admirable work but making wild birds put on a show is not part of it).

When I first wrote about swans in a book called *Waters of the Wild Swan* (Cape, 1992), I attempted to articulate my fascination for whooper swans, in particular:

> *I see in whooper swans a kind of kindred spirit in nature, a restless haunter of lonely places, an inclination to the northlands of the world and a marked determination to preserve its wilderness instincts and keep a wary distance from man and his works…*

What happens at Caerlaverock is a betrayal of that idea, at the very least.

In all those landscapes, I looked for a family group with six cygnets. It was frankly daft, because even if I found one (I didn't – several of four, one of five), there was no way of knowing whence they had flown, but I rationalised it by convincing myself that it was just because I was there in Lofoten that week, when summer in the Arctic took leave of its senses, and I saw what I saw and my heart turned over.

◉ ◉ ◉

I stopped working after I had been watching the Lofoten swans for a while. That is, I stopped consciously noticing details, the actions and shapes of the birds as they swam and fed and lazed and preened, the light on the water as the sun moved round above the mountain, the varying shades of green surrounding the lake and climbing high into mountain gullies, the way the mountain's aura seemed to cover the lake in a kind of protective haze of diffused light that had the effect of binding mountain and lake closer together so that the lake was more obviously a part of the mountain rather than a part of the island beyond the mountain.

In a way, then, the watching became disengaged from swan and water and mountain and sky and was replaced by an over-arching absorption in the complete landscape, almost as if I had *become* landscape. I have met the feeling before from time to time, but always in landscapes with which I have a long and intimate connection. Yet this was still my first day exploring Lofoten, and as such the connection was unique. Whatever I may or may not take away from the moment and from my scrutiny of the landscape and its creatures, the quality of *seeing* becomes extraordinary and utterly memorable. It permits, for example, an imaginary exploration not just of what I had seen the diver do, or the sea eagle and ravens or the swan family on their lake, but all their flights and everyday movements and events throughout their relationship with the mountain, and all the mountain's moods outwith this extraordinary heatwave, what Nan Shepherd called "the total mountain".

It brought to mind a painting I have by an Edinburgh-based artist called Frances Thwaites (1908–1987), a contemporary and collaborator of Anne Redpath and Elizabeth Blackadder among others. Frankie Thwaites, as she was known, was a friend and colleague of George Garson (1930–2010), mosaicist and stained-glass artist of distinction and the greatest friend I ever had. George had been her literary executor and curated a retrospective of her work when she died. He took me to see it. I was intrigued, I had never seen anything quite like it, and he was so pleased by my response that he gave me the painting in question. It is one of a series and called "Flight (No.7)". It shows a pale abstract landscape crossed by many dark diagonals and one or two horizontals. Every line represents a flightpath, the routes followed by birds or butterflies or Boeings when they cross a particular landscape. Some lines are bold and broad and close up to the viewer, others are frail and distant. Some curve and some don't. All are unwavering. The effect is to create a profound depth of field, to achieve great *distance*, and it set my mind to wondering about the nature of all the fliers. That thick horizontal advancing from the left-hand edge of the canvas deep into the picture then stopping abruptly – was that an aircraft entering a cloud, or a family of swans landing on a lake? You don't see what flew, merely the route they took for as long as they were within the force-field of the landscape. The rest is up to you.

George had reproached her about the one aspect of the "Flight" series of paintings that irritated him: a few tiny spots of white, scattered at random. He told me he stopped being irritated by them when she explained.

"The white spots," she said, "are infinity."

One of the diagonals on the painting I have is thicker and bolder than all the others. I decided it was my flight-path, my travels across the landscapes of my own life, and I wondered if Frankie Thwaites thought of it as hers when she painted it. Again, in *Waters of the Wild Swan,* I had expressed a thought that in a way had propelled a flightpath of my own from that very day when I wrote it down sometime in 1991 to that day twenty-seven years later when I stood by the lake under Blåtinden:

> *...Often during the long weeks of watching the mute swans under the mountain on the Highland Edge, my flightpath would wander in its mind at least to the whooper nesting grounds...and I would find that prospect somehow more appropriate for swans. It remains a prospect, however, one to put to the test some other spring and summer, but still, my idea of what it would be like has only forged a closer bond with the whooper tribe, only strengthened a resolve to know it on its native heath. A handful have, after all, got to know me on mine...*

I have just read that for the first time in quite a few years and with a photograph beside me of exactly what it is like on the nesting ground of one particular swan family in Lofoten, a souvenir of perhaps the first leg of a new flightpath by which I have begun to know them on their native heath. The thing is, it may prove to be all I need, because how, I am inclined to ask myself, can I improve on that?

◉ ◉ ◉

A year later, looking at a map of the Lofoten island of Vestvågøya, retracing travels with my finger and lingering over some of the names scattered through the landscape (I speak no Norwegian at all but I am always curious about how a landscape is named and why) I paused at Keilvatnet, the lake under Blåtinden where I had seen the swans. By this time I knew that *vatnet* was simply "lake", but what was *keil*? Bear in mind that I had imagined the sense of a protective screen of diffused light during the time I watched the swans and their lake. I made a rudimentary inquiry about the meaning and the nearest thing I could come up with was *kell* and was astounded by the definition of "that which is covered over". It's a long shot, of course, but perhaps two of us had paused here in similar conditions several hundred years apart: someone who lived here or at least paused long enough to want to name the lake, and I coming curiously and in a spirit of openness towards what to me is a new land, and we had both seen or felt the presence of a kind of protecting veil by which mountain and lake and swans and travellers were bound. I felt strangely involved, mysteriously drawn to the place, although I was a thousand miles away again.

There was be one more resonance, and that too would happen a year after the event, when I was recalling to mind that mountain, that lake, those swans, those divers. I was refreshing my memory about the mountain itself, reading notes I had made and looking at photographs I had taken, then a vast array of online images. And all the time, it was reminding me of somewhere else. I had been vaguely aware of it at the time, but now the awareness had become acute, so that it started to bother me. It was not just a degree of familiarity about the mountain profile, it was everything – the location on an island shore, the lake that seemed to belong to it, and the specific birds. I looked at the name of the mountain on the page: Blåtinden. Again I looked up the meaning of its component parts. *Tinde* is common enough and just means "mountain peak"; *Blå* is easy, too – "blue". My train of thought had been freighted with the possibilities of two mountains with a strong physical resemblance that were both central to encounters with those specific birds. And then it all clicked into place. It clicked into place because they also have the same name.

In Lofoten, the name is pure Norse: Blue Mountain Peak. On the Isle of Skye, off Scotland's west coast, an island to which I have travelled addictively for more than forty years, the mountain name is only part Norse, a relic from the Viking occupation like so many Hebridean names; the other part is Gaelic: Blà Bheinn. Blue Mountain. And if pressed to compile a shortlist of the mountains that have endeared themselves to me most powerfully, Blà Bheinn would be as near to the top of such a list as makes no difference. I have known it for more than half my life. Blåtinden I have known for that single summer encounter, and never set foot on it. But it burns its image and its name and its birds and its lake into my mind with such intensity that it can summon Blà Bheinn, and that is just one of many reasons why the tenacity of Lofoten's embrace has claimed a unique hold on my life.

Chapter Twelve

The Land of Havørn (2) – Islands of Dreams

MY FIRST THOUGHT ABOUT the unprepossessing little hill called Palheia was that this could have been what Scotland looked like 500 years ago. It is not, strictly speaking, Lofoten, but on the island of Hadseløya in the Vesterålen archipelago a little to the north. From its all-but-sea-level mountain birchwood to its 1,400-feet summit carpeted with alpine lady's-mantle, Palheia is rooted in my mind as a kind of wilderness garden of my dreams. It was also a startling manifestation – to my eyes, at least – of what a hillside in the Arctic can look like in high summer.

It wasn't just that I love mountain birchwoods, or that this one clambered uphill looking airily light-footed, it was also that it reached a natural treeline where it thinned and diminished and evolved into montane scrub. In Scotland, thanks to a combination of over-zealously manicured grouse moor and over-zealously neglected deer forest, a natural treeline and montane scrub are as thin on the ground as dodos. On Palheia, juniper thickened the spaces between and beneath the birches with shrubby, knee-high-to-waist-high deep green clouds.

The shelter offered by the combination of birch and

juniper, a mild, wet spring and early summer followed by this blast of unearthly, un-Arctic heat, flung unprecedented growths of flowers all across the hillside, so eye-catching and unexpected that I scribbled them down, not so much to make a list (which would be singularly uncharacteristic of me), more in the spirit of a visual poem: clover, buttercup, bluebell, Arctic harebell, lupin, cranesbill, dwarf cornel, cloudberry, twinflower (purely a pinewood specialist in Highland Scotland and never this profuse), heath-spotted orchid, small white orchid, moss campion and alpine lady's-mantle. Lower down the hill many of these overlapped gleefully, creating the kind of interloping, seeping patchwork you can contrive with watercolours, but higher up, a few established their own zones. In ascending order these were the dwarf cornel zone, the cloudberry zone, the heather zone then finally the summit with its cap of alpine lady's-mantle, so dense it that it could only be avoided by stepping on protruding patches of bare rock. The flower of the alpine lady's-mantle is indifferent to the point of boring, like something withered on a stick, but a light mountain breeze riffling through such a thicket of it reveals the undersides and the rims of the leaves and these are silver, gloriously so in such oceanic sunlight.

A peregrine called from a crag, that hacking, back-of-the-throat cough of a cry. The peregrine is blessed by nature with special gifts, notably a beauty of design, grace and speed of flight, but it is tempting to think that nature was having a laugh when it gave the falcon that voice. My second Norwegian sea eagle in two days emerged from nowhere at all in the middle of the sky just above the last of the trees (how does it do that, appear in such acres of space and waving its great banner of wingspan at you, and yet you never saw it arrive?).

Beyond the trees, climbing beyond the cloudberry zone onto the sudden familiarity of a hillside clothed in short, wind-scoured heather, I had an idle thought: is it possible that there are golden plover in a place like this, a landscape that suddenly so resembled their nesting habitat in Scotland? With the thought unresolved, the summit arrived with a view north to sharp-pointed, sharp-ridged mountains and unbroken slopes that fell steeply to the sea, or rather to the skinny strip of flat and marginally cultivable land where the human natives lived. The built communities throughout these islands give the impression of having been washed ashore along such coastal strips or at the heads of voes, first shipwrecked then tidied up and painted and made orderly and presentable, often beautifully so, but everywhere the mountains leave the people very little room for manoeuvre.

That northwards view looked like an old film from the early days of technicolour: the sky the bluest of blues, the sea only a marginally darker shade of the same, the mountainside that rose with crafted elegance to its peak was vivid green to the very edge of the ridge, the very point of the summit, and only screes interrupted the astounding flow of flowers and grasses. If you had asked me what I had expected of the Arctic Circle archipelago on the edge of the Norwegian Sea before I got here, this would not have been close to my answer.

Something was happening on that distant peak, for it seemed suddenly to have splintered. The glasses revealed three sea eagles that had clearly been perched together around the summit, and now they lifted close together and in unison. What had caught my eye was the chaotic overlapping of six giant wings, hugely flexing high and low, hugely curving and straightening as the three birds heaved into the air and headed for the ridge. Two white-tailed, grey-headed adults were easy to identify in such sunlight, and the darker third bird was presumably the newly flown eaglet. Some chick. Five sea eagles in two days, and not one of the sightings dependent on a sea eagle safari boat loaded with fish to lure them close, although Norway also offers many of those. But really, all you need to do is go for a walk and look up. The three birds from the summit soon fell into their giant stride, taking the line of the ridge, buoyed by thermals, taking it easy, the young bird noticeably less fluent in flight than the adults, falling behind, driving forward to catch up, then all three side-slipped off the ridge out of sight beyond its far side.

These encounters would become the norm. There would be sea eagles every day of the week, binding into place my sense of these islands as the idealised source of Scotland's sea eagles, and nourishing my hopes that one day soon, Scotland's island west will be a sea-eagles-every-day kind of landscape, too, and that from that stronghold they reach out to every corner of my country, island and mainland, Highland and Lowland, for that has been their way in Norway.

Retracing my steps down this idyllic little mountain, slowing the pace, lingering, drinking in the place, writing it down, stopping to stand and stare, to sit and stare, finally pausing for a drink and a protracted seat on a warm granite slab, I was filling my head and my heart with the wonders of the place and the cup-runneth-over day, when the mountain spoke softly to me. Twice. The voice it spoke with has all the tonal quality of a penny whistle. If you know the instrument, you will know that you can increase the pitch by a semitone – or any fraction of a tone – by lifting your finger partially from a closed hole. The effect more or less perfectly mimics the call of a golden plover. There were two birds about five yards apart and fifty yards away from the slab, and more or less exactly where I had been when I had pondered the possibility of their presence earlier in the day. Nice when that happens! I now pondered the possibility that these two birds were close to their northmost limit on the planet. I had read in Seton Gordon's *Amid Snowy Wastes* a chapter on the presence of the purple sandpiper in Spitzbergen. He wrote:

> *All of the larger waders are absent – golden plover, grey plover, curlew, whimbrel, oystercatcher, redshank, greenshank; even the hardy dotterel, which nests on the high Scottish hills does not penetrate thus far north, for suitable feeding grounds for the larger wading birds appear to be non-existent.*

I wondered if climate change would take care of that. Perhaps it already had, and these were pioneer plovers. But as they are among the most totemic birds of wild country to which their voice seems so peculiarly well attuned, I was very pleased to hear them, and for a moment, even in that most precious of landscapes, I felt a pang for the wild places of home – the eternal affliction of the Scot abroad!

◉ ◉ ◉

The entire week was as crammed with incident and encounter as Lofoten is crammed with mountains. They tumble over themselves as I cast among souvenirs in the form of photographs and notebooks and conversations. For a nature writer travelling north, my preferred direction, and tasting new flavours of my very idea of northness, Lofoten was a feast.

The ferry to Skrova, a small island off Austvågøya, largest of the Lofoten islands, was another dream-like venture, horizons characterised by tiers of mainland mountains that paled into distance, and still the heatwave held, still the natives were in a kind of daze at it all. It was memorable for a patch of coastal woodland thick with rowans (redwings on the berries) and a meeting at a distance of about 100 yards with a family group of four sea eagles on top of a low sea cliff. The four rose, again with that chaotic tumult of wings and briefly filling the binoculars, each bird at a different angle, and against a background of that same mountain-strewn sea; an image I will not expect to be eclipsed if I go on eagle watching until I am a hundred. The adult male turned and lingered, and began circling directly overhead, apparently just as intrigued about the human presence. He made twenty slow circles there (you wouldn't catch a golden eagle doing that). Turning to watch him in the glasses for every one of those circles, becoming faintly dizzy, and watching intently how he used his wings and tail to achieve tight turns and wide turns, small fluctuations in height, gliding and soaring (his final gesture, he went straight up with wings battering the air in loose curves then headed out to sea), I felt as if I had been given a personal masterclass in the wiles of sea eagle flight.

Finally, there was the old wooden fishing village of Nusfjord, where accommodation for the second half of the week was in a fisherman's cabin (wonderfully Ikea-ed into the twenty-first century) built on wooden piles out beyond the high-water mark so that the tide flowing into a short, sheltering fjord came in beneath the cabin. Hefty clusters of mussel shells clung to the supporting piles, and tysties nested under the floor. The sound of the sea beneath the bed eased me into sleep. I was never happier inside four walls in my entire life.

Nusfjord is a village whose summers are sung to a soundtrack of kittiwakes. A concession to modern living was an area of decking in front of the cabin, and the perfect place from which to watch the kittiwakes (the Norwegian word is *krykkje*, so that's two Norwegian bird names I know: it's a start) that nested on the cliff across the fjord, about 200 yards away. It will be clear by now that kittiwakes and I go back a long way and that they are among my most enduring and endearing manifestations of summer. Now they were my nearest neighbours in the house of my dreams.

But it was quickly apparent that something was wrong. First of all, there were no more than a dozen nests and faded evidence from empty, whitened ledges suggested that in old summers there had been several times that number. But it was July 22 and there were no young at all. At the mouth of the fjord was the same evidence on the cliff faces: two more kittiwake colonies, both of them completely abandoned.

Yet still the village resounded to the cries of, "Kitt-ay-wake! Kitt–ay-wake!" Something very strange had happened. Kittiwakes had moved into the village. They nested IN the village, ON the timber buildings, on the window ledges and architectural nooks of the general store and the souvenir shop and the oldest houses, and these nests had young. I counted eight twin-chick nests. And two fully-fledged chicks perched side by side on a corrugated roof away from the nests. Nusfjord is a tourist village. Its streets throng with visitors pointing cameras and being loud. It seemed to me the parent birds were distressed, and yet they put up with it rather than using the cliffs a few hundred yards away. Nor did they appear to use their last-resort technique when they feel threatened, which is to throw up a vile liquid at their aggressors, from skuas to sea eagles. If you get it on your clothes, my best advice is to burn them.

Questions. Do the street urchin chicks pass on the habit to the next generation? Will the cliff-nesting habit disappear from this corner of Lofoten? Will the building-nesting habit spread? Will it surface somewhere else? Or will the well-known crisis afflicting kittiwakes all across their range because of food shortage solve the villagers' problem with the sledgehammer of extinction?

Out at the cabin, late in the evening, the sound of kittiwakes still resounds. The relentless daylight drifts past midnight, past 1a.m., and only in the brief twilight towards 2a.m. did I suddenly realise that the voices had stopped. The uncanny silence lasted an hour.

The Arctic has been central to our awareness of the chaotically out-of-kilter climate from the first. The climatologists who first flagged up warming seas and dwindling ice sheets and melting glaciers warned us that what begins in the far north of the world does not stay there.

The process has progressed from there to what we face today: wildfires from the Americas to Australia; the UK's and Europe's hottest ever temperature recorded in the summer of 2019; temperatures in Australia in their summer of 2019–20 edging up towards fifty degrees; polar ice – at both poles – melting at a rate that threatens to reach tipping point, beyond which we can only try to slow down the thawing process but not reverse it.

And in a week of extraordinary beauty in the Lofoten Islands off north-west Norway, a Scottish nature writer with a lifelong thirst for the north of his country and the north of the world, finally crossed inside the Arctic Circle and an airline pilot announced the temperature on the ground was thirty-four degrees, and the first Norwegian he spoke to was the car hire man and his first words were:

"Hi, welcome to Spain."

Chapter Thirteen

The Climate Imperative

WHEN SETON GORDON wrote *Amid Snowy Wastes,* he provided a rare snapshot of high summer in the high Arctic 100 years ago. It was almost the first thing I read when I came back from my week in Lofoten, for I was eager to refresh my memory of his reading of that Arctic landscape back then, while carefully bearing in mind of course that he was fully 400 miles further north. I found this in a chapter titled "Summer Weather in Spitzbergen":

> *The finest month of the whole year is usually June, when continuous sunshine may be experienced day and night, for a week on end; but in June the winter snow still covers the ground to sea level. Indeed, in the Green Harbour inlet it was practically unbroken and lay several feet deep on June 2...Even in the finest summer weather in Spitzbergen it is still chilly out of the sun. At about six o'clock on the morning at Gipps Bay, and not more than fifty feet above the sea, there was black ice on the pools in the shade...Perhaps the most bitter day I remember was June 30. Three companions and I had been floundering all through the night in horrible snow bogs. I shall never forget the bitter wind in which we cowered on the shelterless shore of Richard Lagoon while we awaited a boat from the sloop, anchored over a mile from the shore. It was fully as cold as a January day in Scotland...*

In the summer of 2019, Sami reindeer herders made an unprecedented plea to the Swedish government. A combination of protracted drought and wildfires inside the Arctic Circle had created a situation that suddenly became symptomatic of the impact that climate chaos was making in the most unlikely of circumstances. The Sami were seeking emergency funds to provide supplementary fodder because drought and fire had effectively destroyed reindeer winter grazing grounds. By their own reckoning (and few people are so attuned to the intimacies of the Arctic landscape or better placed to make such a judgment), it could take thirty years to restore those wintering grounds. That presupposes that conditions will prevail that permit the land to heal. If drought and wildfire become the summer norm, then the nature of summer in the high Arctic could become very ugly indeed, and a centuries-old way of life is at risk. Climate chaos has also infected winter conditions for reindeer, for rain has replaced snow in many places. Rain at such latitudes creates an ice sheet rather than a soft fleece of snow, and reindeer cannot reach the lichen beneath. The Sami fear that without government help, the reindeer will starve, and given that reindeer are central to the Sami's way of life, then that too is at risk.

One of the consequences of writing this series of books is that for the past five years I have been locked into a detailed scrutiny of the seasons. And the chaotic nature of what our seasons have become has imposed itself on all the books, even as I was bearing witness to the disintegration of our idea of a wild year divided into four distinct seasons of more or less similar duration. It began with *The Nature of Autumn* and a storm, the first of those storms that the Met Office now likes to call by their Christian names, Storm Abigail. One of the core principles of the way that I work is to spend time being still in nature's company, to listen to the land. Abigail quickly skipped through the polite yellow and amber warnings to red, which is Met Office language for "prepare to meet thy doom". She dumped tons of snow on the mountains, gales reached ninety miles per hour in the islands and 120 in the Cairngorms, and it rained as if all that was left to us all was to build arks. I spent some time along the Highland Edge tasting all this, and came away with the sense that bad as it was, it was no more than a tiny symptom of something much larger, something oceanic, even global. I wrote in *The Nature of Autumn* that "nature is restless", and I described walking the next day among the familiar landscapes of the Highland Edge and finding the place seething with a kind of wild energy; that it felt like a phenomenon of the land itself:

> *I don't believe that the land is neutral. I do believe that it reaches out to us, that it is capable of language couched in terms we might understand if we are prepared to listen. It offers guidance, a better way of coexistence between it and ourselves, because right now our relationship with it is not in the interests of either the land or ourselves.*

That evening, the evening of the day after the worst of Storm Abigail, I heard for the first time about a study of a huge glacier in Greenland called Zachariæ Isstrøm, and it is melting…melting at the rate of five billion tons a year. It was the first time I had been scared by a number. It would not be the last.

A series of seasons followed that could only really be defined by the dates of the calendar, rather than the weather itself. This, I imagine, is why the Met Office started to announce on its weather forecasts that "tomorrow is the first day of meteorological winter/spring, etc." The habit has made it look foolish on a fairly regular basis, most noticeably when it announced in 2018 "tomorrow is the first day of meteorological spring" and the Beast from the East arrived and promptly dumped eighteen inches of snow in my back garden, which was eighteen inches more than the previous winter had managed. The summer of 2019 lasted little more than a fortnight in my neck of the woods (yet contrived a temperature of thirty degrees, while the south of England and France achieved unprecedented highs in such a slim window of opportunity), and the world began to catch fire in a worrying, headline-grabbing way; it hasn't really stopped burning since in lands as diverse as the Arctic, California, Bolivia and Australia. By the turn of the year, Australia's summer achieved temperatures above forty degrees in every state in the country, and exacerbated by drought and strong winds, the wildfires grew vast and ferocious, as potent a warning to the whole world that climate chaos is no longer a nightmare: it's here and it's now.

The summer of 2019 was also notable (at least to a Scottish nature writer who had started to be scared by numbers) for the number of scientific reports that devoured an unprecedented acreage of column inches in newspapers and scientific journals, hours of news broadcasts. They were about the impact of climate chaos on our planet and our wildlife. I was a newspaper journalist in a previous life and have never broken the newspaper habit, nor the habit of amassing a temporary database by amassing a small hillock of cuttings:

A DYING OCEAN: THE ARCTIC AS WE KNOW IT IS ABOUT TO BECOME HISTORY

THE GLACIERS OF ICELAND SEEMED ETERNAL: ONE BY ONE THEY MELT

DRASTIC CHANGES IN RIVER FLOODING AND SPRING GROWING PATTERNS

GREENLAND'S MELTING GLACIERS: 'THAT NOISE YOU HEAR – THAT NOISE IS THE END OF THE WORLD'

ANTARCTIC SEA ICE PLUNGES IN DRAMATIC REVERSAL OF 40-YEAR TREND

MELTING OF HIMALYAN GLACIERS DOUBLES IN 20 YEARS

IMPACT OF GLOBAL HEATING ON LAND POSES THREAT TO CIVILISATION – UN

MIGRATING GEESE SHIFT NORTHWARDS IN RESPONSE TO GLOBAL HEATING

NATURE WARNING SPARKS ACTIVISTS' CALL OVER FOOD

A LAST CHANCE FOR THE OCEAN

These are the cuttings uppermost on my table as I write, what you might call the tip of the iceberg.

I could start with the small numbers. The Arctic has heated up by an average of two degrees centigrade since the pre-industrial era, twice the average for the whole world, and in some places by four degrees centigrade (the troublingly termed "hotspots", and really the Arctic shouldn't have any of those). Sea ice melts earlier in spring, freezes later in autumn, gets thinner and recedes further in summer. So instead of summer ice that reflects sunlight back into the atmosphere, there is a summer ocean which absorbs the sun's heat and grows warmer, and warmer, and warmer.

Now, here come the scary numbers. Since 1979, when we began to be able to measure such things by satellite, Arctic sea ice has lost forty per cent of its reach and seventy per cent of its volume. Science's best guess: it diminishes at the rate of 10,000 tons *every second*. If the trend remains unchecked, and there is nothing about the behaviour of the world's most influential nations to suggest there is any concerted willingness to impose the necessary drastic measures, ice-free Arctic summers could be the norm in twenty years, which means cruise ships all the way to the North Pole, commercial exploitation, Arctic tourism…or to put it another way: "The Arctic as we know it is about to become history."

⦿ ⦿ ⦿

The newspaper cuttings lie untidily on the table beside me as I write. There is no tidy way to accommodate a pile of newspaper cuttings. For a start, quite a few of them are torn out, so at least one of the edges is ragged. Some are cut so are less than the size of a full page, so they slide awkwardly among the full pages. Some are entire two-page spreads with lots of lines untidily marked in ink.

When I started to write this chapter I moved the pile from the far end of the table to lie beside my laptop. I did not put them down carefully, because I had a coffee cup in the other hand, and they landed awkwardly, partly on a small stack of notebooks. Very slowly, they started to slide, and as they slid they reached the edge of the table and fell on the floor. But they are sheets of newspaper of different sizes and different weights, and as they fell they took on the guise of a slow-motion waterfall, or – the thought struck me at precisely the pertinent moment – a glacier. And on the top of the pile, and held aloft and almost but not quite motionless as the mass of the others slid away beneath it, was the one with the headline that read:

"The glaciers of Iceland seemed eternal: One by one they melt"

I played with the words in my head: "The cuttings of ice lands seemed ephemeral. One by one, they fall." It's the sort of thing writers get up to (some writers, at least; okay, this one writer) when they are temporarily at a loss in the matter of writing something useful for the job in hand. The page with that headline was the last one to fall, although it had been on the top. Just like a melting glacier, I thought, the top is the last bit to go.

The area of the table where the glacier of papers had been was now just a piece of oak table, roughly the same shade of brown as the aerial photograph of what was left of the glacier that sat just above the headline on the cutting. So, suddenly I was preoccupied by glaciers. All the other articles were factual pieces written by environment editors and environment correspondents. This one was on the top because it was quite different from the others. It was a singularly moving piece by an Icelandic writer, Andri Snær Magnason, who in his own words had "grown up with glaciers as a geological given, a symbol of eternity". He had been asked by academics in Houston, Texas to write the text for a plaque to commemorate "the first dead glacier in Iceland". Surely no writer was ever handed a more portentous commission.

As it happens, the only time that I have been on a glacier was in Iceland, in the early summer of 1994 when I was making that wretched TV film about swans. To see a glacier, to stand beside one, to touch one, all for the very first time, to watch its inexorable creep from volcanic ice cap to the ocean, is to be hooked for life. The very word "glacier" has grabbed my attention ever since, another powerful component in that thirst for northness that has been with me forever. I had never realised how vocal glaciers are. They groan, grunt, whistle, squeak, growl; you could swear the thing breathes, that it lives, and you try to grasp that its life is a constant process of moment-by-moment journey in actual-time slow-mo that simply never ends. Except that for one glacier in Iceland, ironically named Ok Glacier, it just did. It has been officially declared as "dead ice".

Would you like some more numbers?

Iceland has 400 glaciers. They cover ten per cent of the island. The thickest glacier ice is 1,000 metres. In 200 years, there could be no glaciers at all.

Mr Magnason's article concludes:

'So on the copper plate to commemorate Ok Glacier, we have written to these loved ones of the future: 'We know what is happening and what needs to be done. Only you know if we did it.'

In the greater scheme of things, Iceland is small. If all its glaciers vanished, that would raise sea levels by a centimetre.

If its rather chunkier North Atlantic neighbour Greenland lost its ice sheet tomorrow then sea levels would rise seven metres, which would mean...thank you and goodnight, world. It won't happen, not tomorrow. But (one more scary number alert) in August 2019, NASA recorded the Greenland ice sheet's biggest loss of volume in a single day: 12.5 billion tons.

And this is not just happening in the north of the world. The author of one more study – of ice melt in Antarctica – remarked that the Arctic had become "a poster child for global warming", then pointed out that between 2014 and 2018 Antarctica had lost as much sea ice as the Arctic has lost in the last thirty-four years. Antarctic ice had been consolidating for forty years, but in 2014 "something flipped", and nobody knows what.

In the Himalaya, glacier-melt has doubled in twenty years. In the highest mountains in the world, lost ice is not being replaced by snow. A billion people in India, Pakistan and China, among a cluster of nations, depend on Himalayan rivers for, well...for everything. Untold numbers of wildlife species are at the mercy of the same cataclysmic outcomes.

Given all of the above, it came as no surprise that in August 2019 the United Nations Intergovernmental Panel on Climate Change concluded that the Earth cannot handle the current demands of humanity. Or to put it another way: "We know what is happening and what needs to be done." Oh, and by the way, a glacier just died.

Wildlife hardly ever gets a mention in my collapsing glacier of newspaper cuttings. Two that did give me pause for thought involved bowhead whales and barnacle geese. The geese, travelling from Britain to Svalbard, an archipelago halfway between the north Norwegian mainland and the North Pole, have abandoned a feeding stop south of the Arctic Circle in favour of one north of the Arctic Circle where their preferred grass is growing earlier, in response to global warming. The report of this comparatively minor adaptation has something of a sting in the tail. The geese benefitted from a better food source, and the population grew, but in their new Arctic surroundings, the geese were suddenly at risk from something new to them: polar bears. Something about "the grass is always greener" comes to mind, but so does John Muir's essential truth about everything being hitched to everything else in the universe.

The bowhead whale was an altogether different problem. For thousands of years, the character of the Arctic Ocean has been determined by a band of cold and all-but-fresh water that lies just under the surface of the ocean, a kind of insulation against the creep of warmer waters from the Atlantic that kept the ice intact. That cold layer is disintegrating. Fast. One more study, this time by the Norwegian Institute of Marine Research, predicts that the water will grow too warm for fish larvae and the links of the food chain will start to unravel and fall apart. Whale populations in the Arctic Ocean are in decline already. The bowhead develops slowly, can live to 200 years old, and reproduces only sporadically. A creature that moves to that kind of timescale cannot, simply cannot, adapt biologically to the speed of change and the biological chaos that our impact on the climate is driving.

It just goes on getting truer and truer: everything is

hitched to everything else in the universe. Unless we reverse the warming of the ocean and the disappearance of ice in all its major configurations all across the Earth, from Arctic to Antarctic and from Greenland and Iceland to the high Himalaya and the Alps...unless we do that quickly, we undermine the very essentials that have maintained planet Earth as a beautiful and healthy place to sustain life in myriad forms. Biodiversity as we know it – or to put it another way, everything under the sun – will stop working. The bottom line, the very bottom line, is this: nature is all there is.

There isn't anything else, no alternative, no second coming of natural forces. We have to listen to what nature is telling us, or there will be no us left to listen. We have to start putting nature's needs on a par with our own. We have to start facing up to our responsibilities to nature, because whether we like it or not, we are nature too. And the way we know that is because nature is all there is.

"We know what is happening and what needs to be done."

We need to stop using fossil fuels, yet coal is being massively mined in – among other places – China, America and Australia.

We need to reduce drastically our carbon emissions.

We need to reduce drastically our energy consumption.

We need to replace huge areas of intensive farmland with billions and billions of trees, and stop felling the forests that we still have.

We need to find something better than plastic to do all the jobs that plastic does.

We need to satisfy the needs of the big species with which we share the planet, the ones that sustain all the lesser creatures.

And it is no good waiting for China and the United States and the other heel-dragging nations of the world. It is not good enough to shrug and say, "What's the point when you see what's happening in China and America?"

The point is to set examples.

Every nation that does care enough to legislate drastically on nature's behalf sets an example. And if enough do it, then a tide will turn, and the heel-draggers will be shamed into action. And when do we need to start doing this? Well, a glacier has already died in Iceland. We are already late. We have a lot of time to make up. And we have no time left at all to change our attitudes.

For example: in early May 2019, the United Nations more or less launched that frantic summer tsunami of scientific studies when it announced to the world, with as much fanfare as its considerable public relations muscle could muster, that global scientific opinion thinks we are about to preside over the extinction of one million species. The fanfare was loud, the PR muscle did its job well, and the world's media sat up and took notice. Front pages and two- or three- or four-page spreads were the order of the day. But then, two days later, a royal baby called Archie was born and the million species vanished, never to return. They may as well have become extinct. One London newspaper alone devoted twenty-three pages to the royal baby. That's what I mean about having no time left to change our attitudes.

My levels of outrage may be higher than average. As I

write this now, seven months later, I still shudder at the implications of the million species, and shudder again at the attention span and the priorities of much of the British media. But then nature is not just what I write about, it's where I live, what I breathe, who I am. And before I started to write books about it in 1988, I was a newspaper journalist and I couldn't help noticing then that the media's approach to nature stories routinely involved more puns than biology. Matters can hardly be said to have improved since then.

I heard it first on the radio. My ears said to me:

"Did she just say *one million* species?"

It quickly became clear that she did. And then there was Lofoten when we were coming in to land 120 miles inside the Arctic Circle and the captain spoke first in Norwegian and then in English and one of the passengers said:

"Did he just say *thirty four* degrees?"

I will remember the summer of 2019 as the one when the numbers didn't add up.

Numbers. They are really starting to scare the hell out of me.

Chapter Fourteen

The Accidental Kingfisher and Other Stories – A Diary of Summer

THE SUMMER OF 2019 turned out to be a furtive creature, the summer that mostly hid its fair face behind mountains of rainclouds, or just stayed away altogether. One of the results was a summer of demented rivers and overflowing lochs. When summer did show up it was as short-lived as snowflakes and I clutched at it as a drowning man clutches at straws, with much the same outcomes. Yet for a couple of weeks it bore down with tormenting heat, and just when we were all lamenting the unprecedented lack of butterflies they all arrived at once and flooded the land, and hung around until November.

And it was the summer when the swifts stayed away, at least they did here. I found them eventually, but in the last place I would ever have thought of looking for them. In fact, I wasn't looking for them at all when I found them, and now that I think about it, I didn't find them. They found me. So one way and another, it was an out-of-sorts summer. And at the very end, it was the summer of the accidental kingfisher.

It was also the summer I bought a new bike. Cycling and I go back to the age of ten. I had a guru, my uncle Stuart

Illingworth, with whom I once cycled along by Hadrian's Wall. My memory of it is hazy now, because it rained so heavily and constantly that it felt as if much of it was underwater. And it was he who introduced me to Lindisfarne when we cycled there in my late teens (see *The Nature of Spring*). He was strictly a touring cyclist, and therefore, so was I. In time I cycled round Mull, I cycled the length of Skye and back, and every yard of its Sleat peninsula again and again. When I lived in Glen Dochart for a few years either side of the millennium, I made a habit of cycling round Loch Tay, or through Glen Orchy and back. But by 2019, I was bikeless, having tried and failed to get on with a mountain bike. In the end I gave it away to one of those organisations that does up old bikes and promotes the many virtues of cycling. It was a long time since I had been completely bikeless and I was occasionally visited by pangs of regret, not about the mountain bike, but about the absence of cycling from my life. Those pangs usually sneaked up on me whenever I saw someone travelling alone through a nice landscape on a nice bike. Then I discovered adventure bikes, a kind of cross between a touring road bike and an off-road bike. It was that simple, and it proved to be an inspired choice. I took the plunge and fell in love with cycling again.

A week of rain relented into a sunny June morning. I pointed the bike north from Callander, a track through the oakwoods beyond Kilmahog and up the west side of Loch Lubnaig, the epicentre of my nature-writing life for more than thirty years. The out-of-sorts summer smiled down on me for all of fifteen minutes then comprehensively soaked me inside one of the blackest clouds I have seen since the eruption of Eyjafjallajökull cast its giant shadow over Iceland and grounded every aircraft in Europe, thereby achieving its own carbon neutrality at a stroke. The rain cloud thrashed up the loch, contriving a bit of a gale out of a flat calm as it came and went. After fifteen minutes more it was gone, the air was calm and smelled of bog myrtle and wild hyacinths. Behind it was summer on a mission. Having soaked me, it set about drying me, and that too was accomplished within about fifteen minutes. It was as if the passing cloud was a cork in the neck of a huge bottle of sunlit heat, and the cork having been popped, the heat poured forth and the land basked. What followed was that summer's finest hour. Suddenly I felt lucky cycling easily along the shore of the loch, and the land wore its most seductive summer sheen, that ultimate, glossy, vivid, summer green of a Highland forest before July-into-August dulls it all and it fails of its own too-much. The light and the rain had rinsed the air clear and clean and supercharged it; nothing was hidden and everything and anything felt possible on such a day.

A wondrous old holly tree stands in a rough and piece-meal and quite possibly accidental hedge by the side of the track. I have a peculiar fascination for hollies that I don't quite understand, or at least I don't understand where it stemmed from because they are never shapely and you could never call one beautiful, and they are dark and shiny and waxy and sharp-edged, but metaphorically speaking at least, they get under my skin in a good way.

There is something "other" about them, and I like that. I can only imagine that this particular holly is old, but I have not the slightest idea how old. I have never seen a tree quite like it, for it appears to have grown round itself and inside itself and into itself, so that branches weave in every direction and the tree wraps itself in hoops and crosses over itself like saltires, and where the branches rub up against each other, which is often, they grow into each other and form bizarre junctions. If Rubik ever designed a tree puzzle to unravel, it would be like this.

The combination of the aftermath of the storm (it had drenched the holly as thoroughly as it had drenched me) and the strong sunlight electrified the sheen on the leaves so that they acquired a hallucinatory edge, rippling surprising shades through that vibrant green. I thought I was investing it with more than was actually there, but I found a surprising ally in Hugh Johnson, whose book *Trees* (Mitchell Beazley, 1973) is my most dependable source whenever I have tree problems:

> *The best-looking ones are in hedgerows where the wind has a pruning effect: shorter shoots keep the foliage dense and stress the contrast of glitter and darkness that is half the pleasure. For this reason, holly is best in summer when the new shoots emerge so soft and shiny that they look (and even feel) wet. The new leaves glow with pink, brown and purple tones as well as green.*

It was such a word-perfect portrait of that trackside holly that I wondered idly if perhaps Hugh Johnson had ever strayed that way and marvelled at that very same tree.

Alistair Scott, a former senior executive in the Forestry Commission, wrote and privately published a remarkable little book in his retirement called *A Pleasure in Scottish Trees*.

In it, he pointed out that the holly "is a useful indicator of grazing pressure...it seems to be common only where grazing animals are restricted"; that its leaves "have one of the highest calorific contents of any tree browsed by animals, and are rich in nutrients"; that "the traditional practice continues of feeding stock and deer on the leaves"; and that "woodmen still refuse to fell holly, just in case". He also identified a small wood of hollies on an island in the Fionn Loch in Torridon that more or less avoids the attention of grazing animals and prospers as a result, and that has put in my mind the idea of a small pilgrimage someday soon to see just what a small wood of hollies on an island in a loch in a such a spectacular West Highland setting might look like.

◉ ◉ ◉

The solitary predetermined purpose of my bike ride (as opposed to the non-specific purpose of drinking in whatever turned up) was to check on the nesting season progress – or lack of it – of a pair of mute swans in a reed bed at the head of the loch. In my back catalogue there is a study of that nest site over more than thirty years that fuelled my addiction to the tribe of wild swans. Books, radio broadcasts, that solitary wretched TV film, and hundreds of newspaper and magazine articles flowed from it, and in all that time, no month passed without at least a couple of visits, and during nesting seasons it was sometimes daily.

My visits are less frequent now, but I still like to see for myself whether the birds have been successful or not. The rate of failure is almost off the scale, so desperate that I have often wondered why they stay loyal to the nest site, for it is prone to flooding, and year after year after year I charted the too-predictable fate of one pair of swans in particular over more than twenty years. When the cob finally died, he was over thirty years old, and his mate would have been well into her twenties when she succumbed. It seemed to me that they broke all the rules all the time – they had very little patience with field guide generalisations about "normal" mute swan habitats and behaviour. I learned from them. I knew them as individuals. They recognised me. I could summon them from quarter of a mile away. I learned not just about them but also about all the creatures with which they shared the head of the loch, the reedbeds, the well-treed river that flows into the loch between them, the oakwoods to the west and the spruce forest to the east and the mountainsides above; everything from otters and foxes and pine martens and red squirrels to little grebes and herons to goldeneye and wintering whooper swans, and from surface-skimming clouds of sand martins and dragonflies to golden and sea eagles. The more I watched the swans the more I learned about their fellow travellers and neighbours and the fly-by-nights who just dropped in and were driven out again, for they were two of the most formidably aggressive swans I have ever seen when it came to defending what they regarded as their domain, the extent of which I revised upwards several times over the years; my final estimate was about half a square mile of water.

So edging north towards the reed beds that June day as the heat built and my clothes steamed, I started to look for a pair of mute swans, and the faint hope (it was always a faint hope, the consequence of low expectation born of so much experience) of a brood of cygnets. They were not hard to find. They stood out of the water on a narrow spit of land over by the riverbank and almost surrounded by floodwater. They were preening. They blazed in the sunlight. The grass where they stood shone. The floodwater threw long reflections of the riverbank trees out into the loch, which was far above its natural shoreline. They were surrounded by about thirty Canada geese. If you knew nothing at all of the history of the place and its swans, you might chance on the scene, spare it a passing glance and find it at least agreeable, perhaps beautiful, perhaps a symbol of nature prospering in high summer. But I knew – from the attitudes of the swans and from the presence of the geese so close to the reed bed that routinely accommodated the swans' nest – that their whole year had been rendered devoid of meaning. The nest was fully submerged by floodwater. There were no cygnets. Again. And June was almost certainly too late for them to start a new nest. I say almost certainly because there was one year when the old cob and pen built five nests in a single season, and a swan nest is a major piece of civil engineering. But that was exceptional.

The situation, or something like it, has played out so often in those last thirty-something years, and much more often than not. To have a trouble-free nesting season, the swans need a gentle and relatively dry April and May; and in the mountainsides of the Highland Edge, there are not many of these. If they had had cygnets, the adults would have had no difficulty herding thirty Canada geese far down the loch, and heaven help any that tried to nest in the riverbank grasses and trees. But as soon as the swans accept – or at least recognise – that the nesting season has failed them again, they drop their territorial guard.

Given how much time and energy I have invested in swans, it is hard not to feel for the birds. It is one of humanity's more likable traits that we mourn the dead and perform the rites of funeral. But one thing of which I am truly certain after all these years is that I don't think like a swan. If they have any sense of mourning at all – and in all the nesting season disasters that I have witnessed at Loch Lubnaig, I have seen no hint of it – it must be as brief as it is undemonstrative.

They will moult, they will feed up, they will defend their portion of the loch against the wintering whoopers that rest up here for a few autumn or winter days or weeks, harass them and drive them as far down the loch as they feel is necessary to maintain their authority. There may be ten or a dozen whoopers, and the mute swan cob will take them all on (the pen may or may not bother to participate; pens tend to be more peaceable) and the whoopers will sail before him, even though they could kill him if they turned on him en masse. But they never do, for swan aggression is almost invariably about territory, and the whoopers have no territorial interest in the winter loch, and they treat the mute swan cob with what looks like grudging respect. They understand territory, too. I saw it at work in Iceland, and was surprised to find that whereas all mute swan disputes are enacted on water, the whoopers resolve them in flight. This year, as so often before, it feels almost as if summer for the mute swans goes on without them.

◉ ◉ ◉

I was far back down the track, almost at the south end of the loch where there is a small Forestry Commission development of timber chalets with a good café, and the café had supplanted the fate of the swans in my mind about half a mile back. All the buildings have eaves, and I was suddenly aware of strange bird activity deep in the shadow of one of the buildings and its surrounding trees. I now saw there was a long puddle at the edge of the track, which was almost all I could make out, that and a series of sporadic white flashes. The bird shapes were dark shadows within dark shadows. The scene had a mysterious restlessness about it, like watching bats in the dusk.

It took a few moments to work out what was going on. Around twenty house martins were flying down from nests in the timber eaves of the chalets to bathe and drink in the puddle, and doubtless to cool off as the day warmed up. I eased the bike as gently towards the puddle as I could but every bird rose at once: a squadron of dark blue and white darts fired into the air, and they all took the same flat-out, tight-cornered route to safety, as if they had rehearsed it many times before – hard right into a space between the café and the trees, hard left round the corner of the café. Dark blue and white never look so good since Denis Law used to wear it.

Did someone say café?

⦾ ⦾ ⦾

A network of narrow farm roads extends west from Stirling, mostly pointing towards the mountains around twenty miles away. It has become a favourite leisurely bike ride, usually a couple of hours respite from a writing shift, ten miles there and back, carrying nothing more than binoculars, camera, sketch pad. These long summer evenings are the best of it, and late June is the longest of the best of it.

This is brown hare country. The thing I like about watching brown hares is their unpredictable nature. There was one sitting back on its hurdies in a field, very much at ease, enjoying the fruits of some of the best grassland in the country. It tipped forward to put its weight on its front feet, ambled forward a yard in a low, curved, head-down-ears-up slouch. Stopped, sat again, ate, filled the binoculars with an over-all coat of deep russet (the evening sunlight giving an illusion of grey-ish-yellow shot through with black).

Repeat.

And again.

Then it came to a bare patch of earth not more than a square yard. Here it lay down, flattened its ears and to all intents and purposes disappeared in plain sight. If it had been lying there when I came along, I wonder if I would have seen it at all. There was no question that the hare was fully aware of me. It stared at me. Eyes of black and gold. A long stillness on the patch of bare earth, matched by a long stillness of my own. I told it softly: "I can do stillness, too."

In the evening quiet, a yellowhammer fizzed on a wire. Ever since I heard someone describe the yellow of yellowhammers as "egg yolk yellow" I have called them eggyolkhammers, which I think of as an improvement, an endearment. It fizzed again. I didn't want to turn to see where it was, somewhere behind me, because I was trying to out-still the brown hare, to see what it did next. It stood, sat back, looked left, looked right, tipped forward again and jogged *towards* me. Unpredictable, see? A few yards from the far side of the field fence, it cut away at an angle to reach the fence perhaps ten yards to the west of me, the way I had been travelling. It wafted through the fence as if no fence existed, and out onto the road, sat back again, stared at me, unblinking stare. This is good. What next?

It turned and started to jog along the road, going west. I eased into the saddle, followed. Fifty yards, a hundred, then it stopped again. Sat. Looked back. Still here, hare.

Turned again and jogged on, a leisurely ten miles an hour. I didn't try and catch up because I have seen a hare do forty miles per hour and Chris Froome doesn't do that, and I was no Chris Froome, even when I was Chris Froome's age, which is not recently. I have no wish to be shown up by a brown hare. The hare used the whole road, ran in curving lines, as though perhaps it was looking for an easy way back into the fields on both sides of the road. Then a tractor materialised a few hundred yards ahead and dispelled that theory. The hare stopped. Stared west at the tractor. I had also stopped when it turned its head to look at me through its left eye. In that position, given the placement of its eyes on the side of its head and its almost 360-degrees vision, I reckon it was watching both me and the tractor at the same time, twin threats from diametrically opposed directions.

It turned through ninety degrees to face a hawthorn hedge. The easier option (as I saw it) was on the other side of the road, a fence I could have got through. The hare ran at the hawthorn hedge, vanished, then reappeared about five seconds later twenty yards inside the field and running. I pulled the bike into the grass verge to let the tractor past, exchanged a wave with the driver. When I looked back into the field beyond the hawthorn hedge I couldn't find the hare. But there were two more eggyolkhammers on the next 100 yards of wire, and they hadn't shifted for the tractor. Or the hare. Or the bike when it passed them, and its rider was looking up at them, smiling.

And then I was looking up smiling at a buzzard. It was a pale phase bird. I have never read a convincing explanation for why the plumage of common buzzards can vary as much as it does. The "standard model" is mid-brown with a pale crescent across its breast. A pale phase bird goes all the way to off-white with a dark brown crescent, or no crescent at all. I pass this way often in the car on my way to Flanders Moss and have got into the habit of slowing down when I drive next to a power line slung from a row of wooden posts, because there is often a buzzard perched there, and not quite as often but more often than not, it will be on the fifth post from the end of the row. Now that I cycle this way as often as I drive, I have started to look out for it more conscientiously. I spotted it from a hundred yards away, assisted by its particularly pale outfit, the thick dark crescent lending its plumage something of the air of an Icelandic sweater. So I slowed right down from 100 yards and the bird stayed put, watching intently. From about fifty yards I started doing my fair impression of a buzzard call. I have called to buzzards for years, to see if I can get a conversation going, or even lure them closer. Mostly they stay silent, but sometimes it works, it works. The bird changed its attitude only to the extent that it started to watch me side-headed, which I took to be an upping of the intensity of its scrutiny. I was almost at a standstill when I passed the buzzard's pole, still calling, but I knew better than to stop and break the spell. It flexed and fidgeted its wings twice at the last moment, readying for take-off, but it stayed put. I had just passed the bird's pole, no more than ten yards from it, when it called back, just once. I felt greeted, acknowledged. I was smiling again.

◉ ◉ ◉

The summer of 2019 was the one when the International Union for the Conservation of Nature put the swift on its Red List, rating it as officially endangered. Weather conditions had held back its arrival from Africa, numbers appeared to be down, and my summer was impoverished.

A few years ago now, sitting in my garden on a late-summer evening with a beer and a book, I became aware of distant swift voices. Something about the nature of the sound, the way it fell on my ears, was different. Swift voices always grab my attention: I simply love to watch them fly and it is that simple. I put down my book and my reading glasses and looked up. The reason why the sound struck me as unusual was apparent at once. They were very, very high, and there were upwards of 200 of them. This I had never seen before. More remarkably, they hung around more or less directly above my garden, for half an hour. I watched until my neck hurt. Then I went back to my beer and my book and my attention drifted back and forward between the words on the page and the soundtrack in the sky.

In a handful of years they have gone from such spectacular gatherings to internationally "endangered", one more casualty of humankind's increasingly cavalier approach to the world we live in and the creatures with which we share it. I am in the habit of walking most mornings to get newspapers and read them over coffee. My walk takes me along a street I have known for more or less forty years. It is an alleyway for swifts. *Was* an alleyway for swifts. Flat-out squadrons of them dived down from the roofs of big old villas and whipped through the airspace of the street, trailing their high-pitched voices behind them like contrails. Sometimes the most moving and uplifting encounters with nature are in the wildest of landscapes and you share the privilege of becoming part of the landscape, part of that wild world that is as vital to a nature writer as oxygen. And sometimes, nature comes in among us, finds something agreeable in our built-up landscapes and for a few midsummer weeks transforms them into something so season-defining that we cheer their arrival and mourn their departure. So it has always been with swifts and me. Better still are the crow-step-and-red-pantiles of steeply tiered Fife coast rooftops in places like St Monans and Pittenweem, when your wandering footsteps will be stopped in their tracks by the sight of twenty swifts against the sea. In the out-of-sorts summer of 2019, I saw no swifts before July 6, when seven emerged in my neighbourhood street, made two circuits of the rooftops, and disappeared, in complete silence.

I don't like the tendency to dismiss the voice of a swift as "screaming". Screaming inevitably implies the outermost edges of heightened emotions. What you hear with swifts is simply conversation; swift-to-swift communication. There will be an evolutionary explanation for why a swift's voice sounds the way it does, and it may well have something to do with cutting through the air during high-speed flight. "Screaming" is just plain wrong.

The day after that silent seven in a Stirling street, I happened to be in Moffat, and headed for one of my favourite cafés, which occupies one end of a wooded garden behind a shop on the main street. A high stone wall flanks one side of the garden and beyond it is a long and narrow cobblestone lane, with terraces of two-storey, slate-roofed cottages on its further side.

I sat down with my coffee and quickly changed the position of my seat because the rooftops and the lane and the sky above were alive with passing showers of swifts. These nineteenth-century slate roofs, with slate-roofed dormers and eaves, are a fixture in the architectural tradition of many Borders towns and villages. There was only one other person in the café garden – as it turned out, a fellow swift fanatic. We lingered there talking swifts and I never spent a more agreeable hour in a café.

Swifts, unlike swift-watchers, don't linger. They are among the latest of migrants to arrive and the earliest to leave. Routinely, they start to move south in early July and are mostly all gone by the middle of August. But that summer, I was back in Moffat on August 24, headed for the same garden café knowing full well that it would be devoid of swifts, only to find that they were still there, thirty or forty of them still making the most of the summer's late heat and its bounty of flying insects.

Another consequence of the chaotic weather pattern that made a mockery of both the summer and the spring that preceded it was a dearth of butterflies. Until July 30, that is, when the painted ladies all turned up at once, and the buddleia bushes of the nation's gardens changed colour. A single buddleia bush, in the same street where the swifts came late and left early and in silence, was where I met the painted ladies. I started to count. I got to fifty-five. It is possible I missed some. It was as if their mass arrival stirred some great lethargy into wakefulness, and suddenly the peacocks and red admirals and the coppers and the northern brown argus and the Scotch argus and tribes and tribes of whites were everywhere. A few of them were still hanging around to enjoy a couple of warm days in November.

Mid-August, the track between Callander and Loch Lubnaig where it follows the River Leny through oakwoods north of Kilmahog, and in strong sunlight it makes for idyllic cycling. It contours the slope through the oak woods so that there are trees above and below, and at the bottom of the slope is the river in turbulent mood after yet more days and days of relentless rain. It flowed black and white and sunlit, and giving voice. It was a roar of massed whispers, orchestral in their range when you really, really listen and pinpoint its bassiest and trebliest voices and its baritones and contraltos. It rises through the deflecting canopies of the oaks so that it reaches you in fragments, yet the fragments cohere as soon as they reach you and they become the whole orchestra, the one river.

I stopped to drink it in from about seventy or eighty feet above, to stand and stare, take photographs, eat blaeberries. The sunlight fell differently on the oaks above and below me. Above the track the sun speared between the tree trunks and through gaps in the canopies in straight lines, broad sunbeams where dragonflies danced. The floor of the wood, with its thick carpet of grass and berries and knee-high hollies and birches and rowans, was splashed with spotlights of yellow where the sun held sway.

But below the path, when I turned through 180 degrees, the effect of the sun on the trees wore an extravagant beauty, lit as only an oak wood knows how to be lit. No wood is better at ingratiating itself with sunlight than a summer oakwood. The steepness of the ground – essentially a Highland river gorge – is part of the magic, the deep green of oak leaves is another, and so is a mature oak tree's insistence on making space for itself.

This piece of country where Highland and Lowland meet is among the most captivating and diverse landscapes in this land or any other. It is also, in my mind, where north and south meet, and once you are this deep into it, this is unmistakably a northern river gorge, a northern oakwood, and there is also no mistaking the Highland character of landscape and light and river.

I was still fourteen days away from the end of summer, and it briefly occurred to me that I could stop right now, right here, in this epiphanic moment, head for my writing base and start to try and make sense of the notebooks, the sketchbooks, the photographs, my memories of this and older summers. But who knows what fourteen days might contribute, with the weather set fair and that particular summer's capacity to delight and disappoint and disgust me in more or less equal measure? And somewhere back there I made a promise to nature to see the final season through to the final hour of the final day, for all that it was ever my least favourite season. And writing this book has taught me, among much else, that really there is no such thing, that nature nurtures and mothers them all equally, as all good mothers do with their children.

I cycled back south between fronds of bracken and other more benevolent ferns, shrubby fronds of birch and blaeberry, and the track was everywhere a tapestry of sunlight and oak shadows, and the river cried up to me with its huge voice of whispers, the anthem of the Highland Edge. I was briefly content, a strange and rare state of mind and body, and once again I flirted with the idea of writing *finis* at the foot of a notebook page. And then the understorey at the very edge of the track exploded. For a moment there was nothing louder than the thrum of tiny wings, and as they burst bluntly up and out from the foliage – startled then panicked by the intrusion of my bike – they flew through the sunlight and shadow of the first two feet of airspace above the track, and they were vivid reddish-brown where they caught the sun and just plain brown where they were shadowed. For perhaps ten seconds they flew ahead of me down the line of the track, but then it seemed to occur to them that if they dived back into the understorey they would be safe from me, whatever I might be, whatever threat I might pose. It was a brood of wrens and there were nine of them. I did some simple arithmetic. Surely it was a third brood, to be so newly fledged in the second half of August. I wished them well and headed home, I was smiling again.

⊚ ⊚ ⊚

The Met Office will have you believe that tomorrow, being the first of September, is the first day of meteorological autumn, which is precisely the point, five years ago, that I embarked on this writing adventure. Nature would seem to have other ideas, as usual. The warm weather, summery in its every nuance, would last a few more days, and while I am inclined to heave a huge sigh of relief when September dawns and I can start to think autumnal thoughts, I was suddenly reluctant to take leave of summer and I decided to hang in there for at least as long as the weather did.

The Leny has a sister river just to the south, a little sister that slips more or less unsung into the Leny's mainstream just to the west of Callander. My kingfisher river. It is not very easy to come close to this lowly water but I know a place and I have cause to love it dearly. In particular, there is a bend in the river where I like to sit, not least because of the view, not least because of the otters and herons and redstarts and goosanders and dippers.

Not least because of the kingfishers.

My seat is on the outside of the bend. On the inside bend of the bank there is a willow tree with seven trunks. Depending on the state of the river, the willow is either perfectly reflected in its dawdling waters, or it fractures into a crazy paving reflection when the river's dander is up. Watch an otter swim by when the river dawdles and see the reflections scatter all through its wake, and then watch the whole upside-down tree magically restore itself over several minutes. It is one of my favourite places to sit and write, and I have tried again and again to draw the willow tree, but never come close to doing it justice. But I'm not done yet.

Beyond the willow and the other trees that line the bank, there is a square mile of floodplain, nothing more than long grass with reedy pools and unguessable peaty depths. Beyond that, the land climbs into foothills, and beyond those the wonderful profile of Ben Ledi, first and last mountain of the Highlands and whose presence has presided over so much of my nature-writing life for so long now. So you will perhaps understand why I hold the place dear.

Late afternoon, early September, and a conviction in me that summer has finally slipped below the horizon, that the light has yellowed beyond anything that summer can contrive: this is autumn's light, and I witnessed the change, sitting there, and I wished them both hail and farewell, for they were the beginning and the end of the greatest adventure of my writing life.

So perhaps what happened next should have been more of an epitaph than the final page of a summer diary, but it seemed to me then that it was appropriate to include it in the diary in order to remind you – and me – about the continuity of nature, that there is no division between the seasons, that they feed into each other seamlessly, that the one begets the next, begets the next, begets the next.

Then the kingfisher.

It came upstream, so from my right, and it was as sun-struck and bejewelled and bedazzling as any kingfisher you ever saw, and it took the bend on the inside and for as long as it takes a kingfisher eye to blink I saw its reflection flare among the reflections of the seven willow trunks and then dowse in the same moment. Because at that particular time of day in that particular season, the tree is not only reflected sevenfold in the river, it also casts seven shadows on the water.

The kingfisher seared on upstream then took the next bend, which is a left-hand bend, also on the inside, so you could say it took the racing line. I always sit here with binoculars close at hand for situations just like this one, and as I followed the kingfisher until it vanished round the next bend, something caught my eye on the bank across the river from the second bend. I focussed on the bank, which for about fifty yards was bare earth and vertical and with tree roots protruding here and there. In other words, it was the perfect place for a kingfisher nest. I had wondered in the past if the local birds might nest there but I had never seen them use it. A kingfisher nest is a messy place. Norman MacCaig caught it nicely:

...it vanishes into its burrow, resplendent
Samurai, returning home
to his stinking slum...

It advertises its presence with a seabird-like splash of droppings, and that was what caught my eye in the binoculars as the kingfisher vanished. It is a precarious site, for the bank is not high and the river has a tendency to flood, but then as our land warms and floods more and more frequently, digging a nest in any riverbank anywhere becomes a chancier business.

I was about a hundred yards away from the nest, and getting closer is not possible without a canoe and I don't have one. Besides, I don't need to get closer. I used my camera's ridiculously long zoom lens to take two pictures, nothing more than reference images, but the white splash is clearly visible.

It was days later before I looked at them on the screen of my laptop. Only then did I realise I had accidentally photographed a kingfisher I hadn't seen. Not only that, it was about fifty yards nearer to me than the nest, so it was very out of focus. And it was in both photographs. I have good eyesight for this kind of thing. There are days when I think I have worked as hand-in-glove with nature as it is possible for me to achieve. And there are days when I'm nowhere, when I blow it, when – for example – I might pass up a very good opportunity to photograph a kingfisher because I was taking a picture of kingfisher shit.

Epilogue

A Daydream of Wolves

I AM INCLINED TO SIT ON after the kingfisher's disappearance, although I have more or less stopped working again. More or less because I never truly stop. Nature is all around us wherever we are so my raw material is never far from me. My dear departed friend, the artist George Garson, whose ghost has had walk-on parts in all four of these books, once told me that sometimes he craved the darkness because only then could he stop looking. We had been sitting in a bar in Edinburgh, a thing we did often. I asked him what he meant. He pointed out an old fellow sitting on a bar stool, side-on to us so we only saw him in profile.

"I'm painting him in my mind," he said. "Look at him. Left elbow on the bar. Chin in his left hand, reading the racing paper, his sideburns are grey but there are wisps of red hair escaping from the back of his cap. And his cap's blue. How often do you see an old man wearing a blue cap? See how the sun hits the lines on his forehead, and what colour would you reach for for his eye? Could be just the side of his head, and his left arm...don't need anything else, maybe a bit of the racing paper."

He was silent for a few moments. Then the old boy lifted his right hand, subconsciously pushed his cap up so that it moved back a bit on his head. George laughed.

"That's better," he said. "That's much better."

But I don't suffer from George's obsessive personality. He told me that himself. He was an alcoholic. Once, when he talked about that, he said that I would never become one because I don't have an obsessive personality. It's true: I haven't become the one and I don't have the other.

And yet...

Now that I have stopped working, sitting by the bend in the river, something of me still keeps watching in some shape or form.

Late afternoon eases quietly into early evening. My eyes wander all over that so-familiar landscape. I think how much I love the mountain. Then it occurs to me that I have never set foot on the floodplain on the other side of the river, and isn't that strange? At that moment exactly, I know I have just finished researching *The Nature of Summer*, that the accidental kingfisher was its final flourish. It's not like the moment when I finish actually writing a book (that is usually a cause for quiet celebration), but the sense of relief is at least comparable. The day is still warm, the river mutters and swishes at my feet, the seven-trunked willow leans far out into midstream and it is as if the whole landscape turns around that singular tree, as if it directs the course of the river, and places the components of its backdrop – floodplain and foothills, forest and mountain – and gives them their cues of light and shade.

As if it conducts the song of that place.

I close my eyes for a few moments.

Then there is a movement in the long grass out on the floodplain.

It is the head – small ears erect, eyes of deep gold – and the back of a grey wolf.

It emerges from the long grass and comes to sit by the willow. Tree and wolf appear to know each other.

The wolf looks up and downstream then straight across the river at me.

Its gaze rests on me. I have known this feeling twice before, to be the centre of a wolf's attention. There is no feeling quite like it on the face of the Earth.

It looks back at the tree. Then it stands and jumps down onto a patch of solid ground that had briefly been a beaver dam two or three years before. It no longer looks like solid ground, for it has grown over with water-loving plants. But the wolf recognises it for what it is and knows it is still solid.

It steps into the current and swims across, heading for a small bank of gravel just below where I sit. From there, it scrambles easily up onto the bank and shakes itself, and I am anointed by river water discarded from the pelt of a wolf. Then it comes up to sit beside me facing the mountain, as I do myself. We are a yard apart, no more.

We share a kind of natural camaraderie, a deep calm. The river is the only sound.

Then, from far off beyond the floodplain, somewhere near the forest under the mountain, a wolf howls.

The wolf by my side looks at the sound, as if it were visible.

I imagine that howl rising out of the trees like a column of smoke.

Smoke signals.

The wolf looks at me. Then it stands and retraces its journey back across to the willow, where it looks directly at the tree, then looks back along its spine, back across the river and finds my eyes again.

Then it simply turns and walks away out into the long grass of the floodplain where it is once again as I had first seen it, just the head and the ears and the top of its back.

It stops once, briefly, tilts its head back so that its muzzle is higher than its ears and howls once towards the mountain.

Answering smoke signals.

It looks back at me again. I stand for a better view and I raise an arm.

Then the wolf turns again and walks off, walks towards the unseen wolf at the foot of the mountain.

A willow with seven trunks on a bend in the river. I always thought there was something special about that.

◉ ◉ ◉

It is September again, and autumn waits its cue from the willow. And September is where I came in, with a childhood memory of geese that began this whole writing adventure, this tetralogy of the seasons. From Greenland to Iceland to Norway and Sweden they have begun again, flying down the northern ocean. There is a deep satisfaction in the knowledge that they will have begun, that the day and the night skies are full of their cries, that the wild swans will be close behind them. It occurs to me now, with five years and four books finally behind me, that as our planet lurches crazily towards the unknown, there is one thing and one thing only that we have to do to stop the chaos, and it is very, very simple. We have to learn to think beyond self.

In Scotland, where I ply my trade, I am as certain as I can ever be that nature would like us to put back the wolf and let its benevolence flow across the land. A symbol – a smoke signal telling the wild world that finally we have begun to think beyond self.

Four in the Morning Moon

Look the four-in-the-morning moon
in the eye
and wonder why
it only ever greets the bruised and broken Earth
with a smile.
Maybe the view from there is blind
to tears and fears
of four-in-the-morning moonwatchers
sees instead only the deep
blue beauty of the world
asleep.

Acknowledgements

MARION CAMPBELL OF KILBERRY in Argyll would have been 100 years old as I write this. Given all that her work and her friendship have meant to me, it is fitting that this book allows me to doff my cap to her one more time in her centenary year. She has been gone these last twenty years but it was thirty years ago when she granted me permission to quote from her work. It feels as if I have been doing it ever since. She appears here in the chapter set in Glen Orchy, an old haunt of mine where I have always felt at ease, and it had a special significance for her, too. I cherish memories of long lunches at the Kilberry Inn and long conversations in castle and cottage. My gratitude that our paths crossed goes on.

Seton Gordon has appeared in the Acknowledgements of all three of the previous books in this tetralogy. He is here again, this time for his writing about the Arctic Circle and the light it shed on my chapters set in the Lofoten Islands of north-west Norway. My thanks once again to James Macdonald Lockhart, Seton Gordon's great-grandson, for permission to quote from *Amid Snowy Wastes*. Seton Gordon is nothing less than the founding father of the modern Scottish nature writing tradition in which I work.

Speaking of Lofoten, I owe a particular debt of gratitude to Sandra Dawn Taylor for her part in making an unforgettable week there possible, and ensuring the smooth running of travel arrangements and much else besides.

Finally, to my publisher Sara Hunt of Saraband, thanks for showing such faith in the work, and in particular, for urging me beyond the modest ambition of a book about autumn to take on the seasons quartet that concludes with this book. Thanks, too, to Craig Hillsley, who has edited all four books both thoughtfully and thoroughly. Short straws don't come much shorter.

And thanks to literary agent Jenny Brown, who, with Sara and Craig, has helped me to grow as a writer and given me the confidence to keep challenging myself. All writers should be this lucky.

JIM CRUMLEY HAS WRITTEN forty books, mostly on the wildlife and wild landscape of his native Scotland. His work has been shortlisted for prestigious awards such as the Wainwright Prize, the Highland Book Prize and the Saltire Society Literary Awards. Jim is a widely published journalist and has a monthly column in *The Scots Magazine*, as well as being a poet and occasional broadcaster on both radio and television.

Also in Jim Crumley's seasons tetralogy

LONGLISTED FOR THE WAINWRIGHT BOOK PRIZE 2017
A pilgrimage through the shapes and shades of autumn

IN AUTUMN NATURE STAGES some of its most enchantingly beautiful displays; yet it's also a period for reflection – melancholy, even – as the days shorten and winter's chill approaches.

Charting the colourful progression from September through October and November, Jim Crumley tells the story of how unfolding autumn affects the wildlife and landscapes of his beloved countryside. Along the way, Jim experiences the deer rut, finds phenomenal redwood trees in the most unexpected of places, and contemplates climate change, the death of his father, and his own love of nature.

He paints an intimate and deeply personal portrait of a moody and majestic season.

"Winter is the anvil on which nature hammers out next spring. Its furnace is cold fire. It fashions motes of life. These endure. Even in the utmost extremes of landscape and weather, they endure."

DURING WINTER, DARK DAYS of wild storms can give way to the perfect, glistening stillness of frost-encrusted winter landscapes – it is the stuff of wonder and beauty, of nature at its utmost. Here, Jim Crumley ventures out to experience first-hand the primitiveness and serenity of nature's rest period.

He bears witness to the lives of remarkable animals – golden eagles, red deer, even whales – as they battle intemperate weather and the turbulence of climate change. And in the snow Jim discovers ancient footsteps that lead him to reflect on his personal nature-writing life – a journey that takes in mountain legends, departed friends and an enduring fascination and deep love for nature. He evokes winter in all its drama, in all its pathos, in all its glory.

A BBC Radio 4 Book of the Week

High in the crag's highest
trees, a mistle thrush sings …

Spring is nature's season of rebirth and rejuvenation. Earth's northern hemisphere tilts towards the sun, winter yields to intensifying light and warmth, and a wild, elemental beauty transforms the Highland landscape and a repertoire of islands from Colonsay to Lindisfarne.

Jim Crumley chronicles the wonder, tumult and spectacle of that transformation, but he shows too that it is no Wordsworthian idyll that unfolds. Climate chaos brings unwanted drama to the lives of badger and fox, seal and seabird and raptor, pine marten and sand martin. Jim lays bare the impact of global warming and urges us all towards a more daring conservation vision that embraces everything from the mountain treeline to a second spring for the wolf.

Also by Jim Crumley

NATURE WRITING

Lakeland Wild
The Nature of Spring
The Nature of Winter
The Nature of Autumn
Nature's Architect
The Eagle's Way
The Great Wood
The Last Wolf
The Winter Whale
Brother Nature
Something Out There
A High and Lonely Place
The Company of Swans
Gulfs of Blue Air
The Heart of the Cairngorms
The Heart of Mull
The Heart of Skye
Among Mountains
Among Islands
Badgers on the Highland Edge
Waters of the Wild Swan
The Pentland Hills
Shetland – Land of the Ocean
Glencoe – Monarch of Glens
West Highland Landscape
St Kilda

ENCOUNTERS IN THE WILD SERIES:

Fox, Barn Owl, Swan, Hare,
Badger, Skylark, Kingfisher, Otter

MEMOIR

The Road and the Miles

FICTION

The Mountain of Light
The Goalie